The Search for Life on Mars

OTHER BOOKS BY HENRY S. F. COOPER, JR.

Apollo on the Moon
Moon Rocks
13: The Flight That Failed
A House in Space

Henry S. F. Cooper, Jr.

The Search for Life on Mars

EVOLUTION

OF AN IDEA

Holt, Rinehart and Winston
New York

Copyright © 1976, 1979, 1980 by Henry S. F. Cooper, Jr.
All rights reserved, including the right to reproduce
this book or portions thereof in any form.
Published by Holt, Rinehart and Winston, 383 Madison Avenue,
New York, New York 10017.
Published simultaneously in Canada by Holt, Rinehart and
Winston of Canada, Limited.

Library of Congress Cataloging in Publication Data
Cooper, Henry S. F.
  The search for life on Mars.
  Includes index.
  1. Life on other planets. 2. Mars (Planet)
I. Title.
QB54.C67   547.999'23   79-20061
ISBN 0-03-046166-9

First Edition

Designer: Amy Hill
Printed in the United States of America
10 9 8 7 6 5 4 3 2 1

Most of this book originally appeared in slightly
different form in The New Yorker.

TO MARY, LIZZIE, HANNAH,
AND ESPECIALLY

Molly,

WHO DIDN'T GET ONE BEFORE.

Acknowledgments

Writing a book can involve almost as many people as sending a spacecraft to Mars. In particular, I want to thank the John Simon Guggenheim Memorial Foundation and, of course, *The New Yorker* magazine for their support; they supplied the rocket fuel. At *The New Yorker*, in whose pages most of the material in this book was initially launched (though in a somewhat different form), I want to thank William Knapp, who edited the first part, which appeared as a profile of Carl Sagan; C. Patrick Crow, who saw the second part, an account of the biology experiments on Mars, into print; and also Sara Spencer and Anne Mortimer-Maddox, who checked the material with the Viking scientists. Collectively, they were the ground crew. I also thank all the scientists mentioned in this book for submitting, first, to periodic interviews, and later, to numerous lengthy phone calls; in particular, both Harold P. Klein, director of life sciences at NASA's Ames Research Center in Mountain View, California, and Norman H. Horowitz, professor of biology at the California

vii

Institute of Technology in Pasadena, gave abundantly of their time—though, of course, they are in no way responsible for any anomalies or glitches. Finally, I want to thank Don Hutter, my editor at Holt, Rinehart and Winston, who is flight director of the present mission—the one that has presently alighted on your hand.

Henry S. F. Cooper, Jr.
New York, N. Y.
November 1979

Contents

ix

Part One

CARL SAGAN

A Resonance with Something Alive

On clear nights, in high, remote areas, Mars is seen to glow with a steady, hard reddish light—something it rarely seems to do from the city, through whose smoggy air the planet looks wan and bloodless and is difficult to identify. Normally, it can be picked out because it is brighter than the stars around it and, unlike them, doesn't twinkle. Through a telescope, Mars is not a pinpoint of light, the way all stars appear through even the most powerful instruments, but a round reddish ball—clearly a place, like our own globe, to which one might travel. In a number of cultures, the red planet has been associated with war, and in the last hundred years it has been the battleground of a scientific one: a dispute over whether life exists there. Of all the planets in the solar system, aside from our own, it has been considered the most likely to harbor living things. To be suitable for life, a planet must clearly be large enough to generate an atmosphere or water internally and to have sufficient gravity to

3

retain it; it must also be the right distance from its sun. Yet no one knows what the specific limits on size or on distance from a sun are. Obviously, the Earth fulfills the conditions, and the Moon does not. What has made Mars particularly tantalizing to the scientists is that in size (it is a little over 4,000 miles in diameter) it is midway between the Earth and the Moon (a little under 8,000 and a little over 2,000 miles in diameter, respectively), and it is only half again as far away from the Sun as the Earth is—on the average, 141 million miles.

During the 1960s the National Aeronautics and Space Administration sent three Mariner spacecraft to observe the planet by flying close by it and, in 1971, a fourth into orbit around it. Between 1962 and 1973, the Soviet Union dispatched eight spacecraft that we know of to Mars; while five sent back some data, none was particularly successful, and three disappeared without a trace. Then, in August and September of 1975, NASA launched two more American craft, Viking 1 and Viking 2—each composed of a lander, which would descend to the surface of Mars, and an orbiter, which would continue to fly above it. They would reach Mars 11 months later, in the summer of 1976, when the planet was on the far side of the Sun (hence invisible in the daytime sky) and approaching its most distant point from the Earth, making communication with spacecraft in its vicinity more difficult. These factors, among others, made many Viking scientists anxious for the success of the missions. Through a curious conjunction of celestial mechanics and more worldly concerns, the first lander was scheduled to touch down on July 4, the day of our Bicentennial.

The most ardent recent advocate of the possibility of life on Mars, and on a lot of other places distant from our planet, has been Carl Sagan, a professor of astronomy at Cornell, who has been on the scientific teams planning several of NASA's unmanned spacecraft missions. On August 10, 1975, the day before Viking 1 was supposed to be launched from the Kennedy Space Center in Cape Canaveral, Florida, Sagan was addressing a dozen or so children seated on the hot cement near the pool of the Ramada Inn at Cocoa Beach, about 12 miles from the launching pad. A youthful-looking man of 41, with long, straight black hair combed at a sloping angle across a high forehead, Sagan is a controversial figure, but most scientists agree that if he didn't embody the spirit of the whole Viking enterprise he at least supplied its imagination. On this occasion, he was dressed in black bathing trunks and a maroon-and-white patterned shirt and was sitting at the edge of the pool. At his feet were a partly broken model of a Viking lander, squat and froglike, and, cradled in a wastebasket, a ruddy-colored globe of Mars as big as a beach ball. The children, most of them under 10, were sons or daughters of Viking scientists or engineers. Sagan's 4-year-old son, Nicholas, was among them. They seemed to like Sagan, a man whose own childhood never seems very far behind him; he has remained close to it and seems to draw from it a rich and playful imagery.

"Here on Earth, we have pools and beach balls and hot dogs and other nice things, but if you were far away you wouldn't see these things," he began. "From space, the Earth would be a blue dot among lots of other dots—blue ones, green ones, brown ones, and, in particular, one big red one." He picked up the globe of Mars. "There is snow up here and down here," he said, touching the poles, where two holes had been punched so that the globe could revolve on a stand. "These holes don't exist on the real planet. But there's a giant mountain here. And here there's another. And here's another. And here's a huge canyon that would stretch from New York to beyond San Francisco if it were on Earth. We know the mountains and the canyon exist because we can see them from space near Mars, but what we don't know is what is down *on* Mars." Though he didn't actually say it, he left the impression that if there actually were not hot dogs on Mars there might well be manifestations of life almost as interesting. One of Sagan's favorite arguments is that if a few thousand years ago, before there were advanced civilizations on Earth, a spaceman from another planet had had a view of the Earth no better than the one we have of Mars, he might not have known that any living thing was here. "So we want to send someone to Mars to see what's there," he went on to the children, who all looked very blond and most of whom had recently come from such places as Hampton, Virginia, or Denver, Colorado—the sites of one or another of the factories, universities, and space centers where work on Viking had been going on. "We thought of sending a Martian, but we didn't know any Martians. That's one reason we're going. We asked some of our friends if they could live on Mars, but none of them could. It's too cold and too dry, and the atmosphere is mostly carbon dioxide instead of oxygen, like ours, and is only about one-hundredth the density of our own. So we had to make a person, and his name is Viking." Sagan picked up the model of the spacecraft. "He's a very special guy, and now I'll tell you what he has. He has three feet; he can't walk on them, but he can bounce a little as he lands. He has one giant ear inside his belly; it couldn't hear you, but it could hear an earthquake—a Marsquake—a thousand miles away. He has two ears on top, which turn, to hear radio signals, and he can talk with them, too. He has two eyes, like ours, only they're on stalks, like a crab's; he can see all the same colors we can and some we can't. Now let's talk about mouths. He has three mouths, one of which is also a nose. With them, he eats dirt."

"*Yuk!*" said a girl with a very blond ponytail dangling down a very bronzed back.

"Yuk for you, but he likes it," Sagan said. "He doesn't eat for energy—he has all the food he needs inside him. He eats because he likes to. He has especially good tasters to tell one kind of dirt from another. He could easily smell the chlorine they're putting in the pool. He could even tell if there was anything *alive* in what he was eating. He has a hand to pick up the dirt he eats. He can pick up other things to look at. He has two arms. One is ten feet long, so he could reach over and pick up that girl over there. The other is shorter, and he uses it to feel the air. It's very thin air. Every day, he radios Earth and says things like, 'Hello, Earth. The temperature on Mars is seventy below zero. It is very windy, and we don't have any snow.' He's pretty smart. He has a vocabulary of eighteen thousand words. An eight-year-old child knows perhaps half that many."

"Would you rather have Viking or an eight-year-old on Mars?" asked a boy in red bathing trunks who looked about that age.

"I'd rather have an eight-year-old," Sagan said at once. "Viking may be smart, but he's slow. If a fat Martian walks by, Dr. Anderson, the scientist in charge of the big ear, will go to Dr. Mutch, the scientist in charge of the two eyes, and say: 'I hear something fat walking around out there.' Then Dr. Mutch will go to Mission Control and say: 'There's something fat walking around out there. Let's look for it.' *Three days later* Viking looks for it, and by that time whatever it was will have lumbered out of view. Another thing that Viking can't do is reproduce. It would be nice if these two Viking landers—there are two of them, remember, each with its own orbiter overhead—could make a lot more, but they can't. Anyway, that's our special guy on Mars, and tomorrow he'll go off. This is the first time something will actually land on Mars and tell us about it, so you're very lucky."

"What if it blows up?" the boy in red trunks asked.

"That's one reason we have two of them," Sagan said.

"What if a Martian cuts off an eye?" another boy asked.

"That will be terrific! Then the other eye will see him."

"What if the Martians have sophisticated weapons that blow it up?" another girl asked.

"Then we'll have a blown-up lander," Sagan said. "And maybe the cameras will photograph the Martians doing these bad things. But the Martians probably won't be bad. They'll either be kindly or they won't care about us."

With his playfulness, his ability to bring science fiction to the aid of science, and his nimble way of turning a question inside out, so that an adverse circumstance suddenly becomes an asset, Sagan alternatively delights and infuriates not only children but his scientific colleagues as well. The latter don't know quite what to make of him, for although they regard him as a good, even brilliant scientist, they have trouble coming to grips with his most distinctive quality, his imagination. Sagan is a theorist—a type of scientist who traditionally irritates many of his fellows because he necessarily deals with what might be instead of what is.

Sagan believes that the most important question facing mankind today is whether there is life, intelligent or not, elsewhere in the universe; although the subject is one that is full of pitfalls, he and a large part of the scientific community have in recent years come to feel that there must be such life. At the time of the Viking mission, there was no direct evidence for extraterrestrial life, although one biologist who is a good friend of Sagan's, Joshua Lederberg, then of Stanford University (and now president of The Rockefeller University), adduced as certain proof that there is life in space one impressive set of evidence: ourselves. There are, however, two generally acknowledged areas of indirect evidence of exobiology. (*Exobiology*, meaning life outside this planet or the study of it, is a word that was coined some 15 years ago by Lederberg, who, in conversations with Sagan and others, found *extraterrestrial biology*, from which it derives, too much of a jawbreaker.) One area of indirect evidence is the vast number of stars in the sky: there are an estimated 250 billion of them in our galaxy alone, and within sight of our largest telescope there are probably as many galaxies as there are stars in this one. Since planets are thought to be a common consequence of star formation, a large proportion of these stars presumably have solar systems, though none except our own star are known to have. This line of reasoning would not prove much unless it could also be shown that life will occur where conditions are right, and the second area of indirect evidence suggests that it will. In the last couple of decades, molecular biologists, who are concerned with the formation of the molecules on which life is based, have demonstrated what they believe is the way life on Earth developed from the simplest organic compounds (organic compounds are those based on carbon). Many of these compounds have recently been discovered in such profusion throughout space that many biologists are convinced that life not only is a common manifestation in the universe but may actually be an inevitable consequence of it. They see life as an extension of the physics and chemistry of a planet. If there is no life elsewhere, Sagan has said—turning the argument back on itself—then scientists will be faced with what he regards as a much more difficult problem: explaining what is so special about our particular part of the universe that life developed only here. Sagan is fond of quoting a friend of his—Philip Morrison, a physicist at the Massachusetts Institute of Technology, who was chairman of a NASA committee to recommend methods of communicating by radio with extraterrestrial civilizations. Morrison has said that the discovery of any sort of life on Mars, however meager, would immediately change life from a miracle into a statistic—initially of two. Indeed, Sagan and others see life in the universe as a sort of statistical pyramid, as it is on Earth, with the lower forms vastly outnumbering the higher; consequently, if a microbe was found on a relatively arid planet like Mars, Sagan feels, many people would be willing to make what he has called "the great leap" to the acceptance of belief in a cosmos populated fairly consistently with intelligent beings.

Sagan has pursued the matter of extraterrestrial life not only in laboratories but also in classrooms, in books, and on television (where he was a familiar sight before and after Viking landed on Mars). He has written 15 books, including 3 popular ones about exobiology: *Intelligent Life in the Universe* (1966), on which he collaborated with the Soviet astrophysicist I. S. Shklovskii; *Other Worlds* (1975); and *The Cosmic Connection* (1973). Later, he would turn to writing about the development of human intelligence and the evolution of the brain, in *The Dragons of Eden* (1977), which won a Pulitzer Prize, and *Broca's Brain* (1979). He would also do a 13-part television series, to be aired in 1980, on the cosmos.

Even when he is writing about the brain, Sagan never strays far from the subject of the universality of life, for which he first became widely known with the publication of *The Cosmic Connection*. It is a literate account of the likelihood of our discovering extraterrestrial life of all sorts, from microbial to intelligent, and its approach ranges from scientific objectivity to lyricism. Sagan writes well. In his preface to *The Cosmic Connection*, he says:

> Even today, there are moments when what I do seems to me like an improbable, if unusually pleasant, dream: To be involved in the exploration of Venus, Mars, Jupiter, and Saturn; to try to duplicate the steps that led to the origin of life four billion years ago on an Earth very different from the one we know; to land instruments on Mars to search there for life; and perhaps to be engaged in a serious effort to communicate with other intelligent beings, if such there be, out there in the dark of the night sky.
>
> Had I been born fifty years earlier, I could have

pursued none of these activities. They were then all figments of the speculative imagination. Had I been born fifty years later, I also could not have been involved in these efforts, except possibly the last, because fifty years from now the preliminary reconnaissance of the solar system, the search for life on Mars, and the study of the origin of life will have been completed. I think myself extraordinarily fortunate to be alive at the one moment in the history of mankind when such ventures are being undertaken.

Not only was each lander an extension of human perception, as Sagan had hinted, but beyond that its designers—confronted with the problem of how to understand a whole new world— had unwittingly created a robot, however rudimentary, that depended very much on the way people think and perceive in this world. And the different elements of this rudimentary human intelligence on Mars had their counterparts in the scientists and scientific teams that were part of the Viking project on Earth and that reacted like the centers of a single brain to what were almost sensory nervous signals from Mars. The entire organism—the teams of scientists on Earth with their mechanical projections on Mars—might even have been thought to possess certain unconscious motives of its own.

There were 13 scientific teams, among them those for Viking's seismometer, meteorology instrument, sampler arm, inorganic and organic chemistry instruments, and the package of three biology instruments. Sagan was a member of what was known as the imaging team for the two Viking landers—the scientific group that would analyze the photographs sent back from the surface of Mars. He and James B. Pollack, of NASA's Ames Research Center in Mountain View, California (Pollack had been a graduate student of Sagan's at Harvard before the latter moved to Cornell) were the only astronomers on that team, which was made up mostly of geologists. More important, Sagan was the only member to have a strong background in biology. Although there was a separate team of biologists, and there were three specialized instruments aboard each lander to detect microbes, Sagan believed that the cameras could well prove the most effective means of discovering life on Mars—on the principle that the surest way to discover life on Earth is to open your eyes. The cameras, he felt, would make the fewest assumptions about what life on Mars would be like; the three biology instruments, in which Martian soil would be cultured to see if anything would grow, would make a number of assumptions about such things as temperature, nutrients, wetness, and metabolism. The cameras would make only one, but that, of course, was a whopper: that life on Mars would be big enough to see. Sagan was about the only Viking scientist who accepted this possibility, and he said at the time that he would make it his principal duty to search the Viking photographs for visible signs of life. It was almost beyond his colleagues' wildest expectations to find a microbe on Mars, let alone anything larger. "Carl serves an important function at some risk to himself," his friend Morrison said at the conclusion of a press conference the day before the scheduled launching of the first Viking spacecraft. (Owing to numerous glitches—aerospace gremlins—it didn't get off the ground for almost 2 weeks after that; in fact, at one point there were so many mechanical problems with Viking that one eminent scientist called it a "screwed-up mess.") At the prelaunch press conference 17 Viking scientists were asked for a show of hands on whether they believed there was life on Mars. At first no hands went up; then 2 or 3 were raised; and after about a minute there were 11. Morrison, who witnessed this demonstration of uncertainty, cited it as evidence that most scientists felt a certain sympathy for Sagan's more open espousal of extraterrestrial life—as though he were their collective unconscious. Most of the scientists in the group, however, thought that the tentativeness of their hand raising expressed their attitude better than Sagan's eloquence did. Most of them were conducting experiments on Mars that were much more prosaic than the task of searching the Viking photographs for visible signs of life, and it was just possible that if Sagan wasn't there to do that, no one else would do it.

Among those who wouldn't devote much time to examining the photographs for life was Bruce C. Murray, a geologist who had been a member, along with Sagan, of the imaging team of Mariner 9, the spacecraft that began orbiting Mars in mid-November of 1971 and, over almost an entire year, sent photographs covering the whole planet. Murray, a thickset, solidly built man, had been a professor of planetary science at the California Institute of Technology; following a stint as leader of the imaging team of Mariner 10, which flew by Venus and Mercury, he was made director of the Jet Propulsion Laboratory, in Pasadena, which Caltech runs for NASA and which is the flight operations center for Viking. (NASA's Langley Research Center, in Hampton, Virginia, is in overall charge of the project.) As a geologist, Murray is fundamentally interested in the Earth, and, unlike Sagan, he has studied the other planets primarily to learn more about this one. Many scientists regard Murray as being, in his own fashion, as impressive as Sagan in his outspokenness. "Bruce is as good in his ultraconservative way as Carl is in his ultra-far-out way," a colleague of both has said. Sagan's ideas about life on Mars irritated Murray. At a symposium on November 12, 1971, the day before Mariner 9 went into orbit around Mars, Murray and Sagan shared a platform at Caltech with Ray Bradbury and Arthur C. Clarke, the science-fiction writers, and Walter Sullivan, the science editor of the *Times*, for a discussion later published as a book, entitled *Mars and the Mind of Man*. Murray, who considered himself the solid dough amid the general leavening, said:

> I really don't think there is any life on Mars. There never has been any evidence of it. It has just been a very attractive idea. You cannot completely disprove the possibility of life there any more than you can disprove life on the moon. . . . It just becomes less and less likely. And it has become very less likely as we have gotten more information about Mars. When you go back to find out why people thought there might be life there, it was in part, if not entirely, the result of this wishful thinking and the Edgar Rice Burroughs kind of popularization.

In Murray's opinion, what had got the study of Mars off on the wrong foot were the ideas of Percival Lowell, a highly imaginative scientist and writer who similarly inspired a host of other writers, including H. G. Wells and Burroughs, the creator of Tarzan of the Apes. (Burroughs, starting in 1917, became one of the most prolific popularizers of Mars, through a series of novels about the miraculous travels there of one John Carter, Virginia gentleman.) Lowell's interest in Mars had begun late in the last century, when he became interested in reports of observations made in 1877 by Giovanni Schiaparelli, an Italian astronomer. Schiaparelli said he had seen faint lines on Mars, and he referred to them as *canali*. The popular British and American interpretation of the word *canali* was that it meant canals—which are, of course, man-made—rather than channels, which need not be. Nor did Schiaparelli make any attempt to clarify the interpretation; indeed, he once remarked; "I am very careful not to combat this suggestion, which contains nothing impossible"—a use of the double negative still favored by seekers after extraterrestrial life, particularly Sagan.

(It is, of course, a not ungrammatical use of the double negative.) Schiaparelli, whose eyesight was failing, continued to observe Mars on its close approaches to the Earth until about 1890. (Mars and the Earth pass each other in their orbits about every 2 years. At these times, they are said to be in opposition; exceptionally close approaches occur about every 16 years.) Lowell, who had exceptionally good eyesight and was proud of it, took up the watch in 1894, when he set up an 18-inch telescope on a hill, which came to be called Mars Hill, outside Flagstaff, Arizona; this was the genesis of the Lowell Observatory, one of the first in this country to be situated in a remote spot for good visibility, and today a major astronomical institution. Then as now, Mars watching had its difficulties. When the planet was low in the sky, so that the telescope's eyepiece was high off the ground, Lowell had to hang from a ladderlike scaffold that lined the observatory walls like the stall bars of a gymnasium. On a drawing board hooked to a convenient rung, Lowell made sketches of his observations. In color, Mars was a brilliant, splotchy orange red, though there were some darkish blue-green splotches as well. In 1659 the Dutch astronomer Christiaan Huygens—the first man to study Mars telescopically—had discovered white splotches at either pole, which he deduced to be polar caps. The markings looked fuzzy and ill-defined, and they varied from time to time, the way a nearsighted man might view the image in a kaleidoscope; Mars had, in fact, the shifty splotchiness of a Rorschach inkblot, in which people are sometimes asked to tell what they think they see.

Over the next 20 years, Lowell concluded that Mars was laced with an elaborate webbing of canals, which, because of their extreme length, precision, and straightness, could have been created only by a highly advanced civilization. Lowell, who was as literate as Sagan (the poet Amy was his sister and his brother Abbott Lawrence was president of Harvard), recorded his observations in three very persuasive books. "Suggestive of a spider's web seen against the grass of a spring morning, a mesh of fine reticulated lines overspreads [the planet]," he wrote in *Mars and Its Canals*, which was published in 1906. "The chief difference between it and a spider's work is one of size, supplemented by greater complexity, but both are joys of geometric beauty. For the lines are of individually uniform width, of exceeding tenuity, and of great length. These are the Martian Canals." Lowell mapped the canals on bone-white globes, which are still displayed in a glass cabinet at the observatory; the cobwebby tracery gives them an antiquated, discarded look. Except for the canals, Mars appeared to Lowell to be mainly a sandy desert—the orange-red parts. The polar ice caps wax and wane alternately with the Martian seasons. Schiaparelli and earlier observers had thought the blue-green splotches were seas; they seemed to grow smaller in cold weather and bigger in warm, as though they were being fed by streams running off that cap whose hemisphere was in its summer phase. Lowell, however, decided that these seasonally changing splotches couldn't be seas because, for one thing, some of the canals ran straight across them. Moreover, the canals themselves seemed to thicken seasonally, as though *they* were carrying water from the poles in warm weather. He was bothered by the fact that to be seen from the Earth at all the canals would have to be implausibly wide. He decided that what he was seeing wasn't water at all but vegetation, such as that which forms broad belts along waterways in dry areas of the Earth; during the warmer months, he reasoned, when water from the polar caps was rushing toward the equator through the canals, the vegetation along the canals' banks, and also in large low-lying areas around them, became lush and more visible. As Lowell himself took delight in growing plants in the Arizona desert, it seemed obvious that the Martians would do the same;

indeed, he felt there was a sort of literary aptness in setting up his telescope in a terrestrial desert because it helped, as he put it, "to explain what the telescope made visible." He was sure that Mars must be an old, dying planet and that the canals were a desperate attempt by the Martians to divide up what water was left. Though these ideas quickly became popular, there were scientists, even in Lowell's day, who did not accept them. One was the Englishman Alfred Russel Wallace, the codiscoverer with Darwin of the theory of natural selection, who wrote of Lowell in 1907:

> The very immensity of this system, and its constant growth and extension during fifteen years of persistent observation, have so completely taken possession of his mind, that, after a very hasty glance at analogous facts and possibilities, he had declared them to be "non-natural"—therefore to be works of art—therefore to necessitate the presence of highly intelligent beings who have designed and constructed them. This idea has colored or governed all his writings on the subject. The innumerable difficulties which it raises have been either ignored, or brushed aside on the flimsiest evidence.

Lowell paid no attention to such slings and arrows. In addition to being superb engineers, he pointed out, the Martians had to possess a wisdom far beyond that of Earthlings because they had to be organized on a planetwide scale to maintain their planetwide waterworks. Men should take note, Lowell concluded, for someday the Earth would reach a similarly desiccated state, and mankind would have to learn to irrigate the Earth with what water was left in our own polar caps. Lowell died in 1916; he is buried a few feet away from his observatory, in a small mausoleum that looks like an observatory itself; its dome is made of a translucent blue glass that lets a sort of eternal twilight fall on the tomb inside. Carved on a wall of the mausoleum is a quotation from one of his books, expressing the hope that "what that other world shall have taught us will redound to a better knowledge of our own, and of that cosmos of which the two form part."

Lowell was by no means the first, nor would he be the last, to find utopias in space, particularly on Mars; Schiaparelli and his successors gave the different splotchy regions of that planet names that had yearning, wistful sounds, such as Arcadia, Elysium, and Eden; there is, in fact, a great plain named Utopia on Mars. Lowell was not a crackpot. Rather, he had developed a theory of extraterrestrial life that he believed in perhaps too deeply; exobiology is a risky field, and its practitioners require a high degree of flexibility. Although Lowell had the misfortune to be wrong, his ideas were enormously influential, and they weren't finally discounted until a decade ago; indeed, they may not have been altogether laid to rest even yet.

To the consternation of Bruce Murray and other scientists, Lowell cast a very long shadow. After his death, planetology went out of fashion, partly because of the claims he had made, but when it became fashionable again, with the advent of the space program in the late 1950s, Lowell's ideas had to be reckoned with once more. Though only a handful of scientists still believed in the canals, a great many still believed that Mars, with its misty Elysiums, might be covered with vegetation. Observations at first seemed to indicate that the atmospheric pressure on the planet's surface was great enough—roughly a tenth of that on Earth—for water to exist there in liquid form. In the late 1940s and early 1950s, spectroscopic equipment indicated the possible presence of water in the polar caps, making it likely that the atmosphere contained some water too. However, in 1965, Mariner 4, the first successful craft to fly by Mars, sent back photographs of about 1 percent of the planet's surface, revealing terrain that not only was devoid of anything resembling canals but was as heavily cratered and as dead-looking as the Moon. Lowell's influence had been so great that up to that time only a few astronomers had suggested the planet would turn out to be cratered by meteors—a fact that would indicate little planetary evolution had taken place. After Mariner 4, the cratering seemed so inevitable to everyone that they could only blame Lowell for their shortsightedness. "It wasn't that Lowell misinterpreted something; he misinterpreted *nothing*," Sagan, who made a point of examining photographs of Mars with Lowell's theories in mind, said later. Like most scientists, Sagan was very much of two minds about Lowell: "He builds a superb observatory, and with it he makes some of the most monumental blunders in scientific history; yet inadvertently his ideas led to many discoveries by others. He predicted the existence of the planet Pluto from some perturbations of the orbit of Uranus, and fourteen years after his death, using his own observatory, others found it just about where he said it would be; yet the perturbations of Uranus's orbit which Lowell used as the basis of his prediction turned out not to be caused by Pluto—a sort of cosmic joke. On another occasion, he had his assistant, V. M. Sipher, try to measure the rotation of the spiral galaxies, which he thought were sort of proto solar systems in our own galaxy. In the course of his investigations, Slipher discovered that they were all receding; this was the underpinning of the theory of the expansion of the universe and the basis of modern cosmology, all from an error of Percival Lowell's. And the third thing he triggered with an enormous mistake has to be the current interest in Mars, for while his

flamboyant ideas turned off the whole contemporary community of astronomers, they turned on all the eight-year-olds who
came after him, and who eventually turned into the present
generation of astronomers." One of these, of course, was Sagan,
who, though he has not made any errors as spectacular as
Lowell's, was occasionally regarded as something of a kindred
spirit.

Exactly what it was that had caused Lowell to see the canals is
not known; if it was a hallucination, it was one shared by a great
many people, some of whom are still living. Sagan thinks it was
an optical illusion; one theory is that the human eye, which
tends to create order out of chaos, might somehow have been
drawing lines between craters that hovered just at the edge of
visibility. Not only was there nothing like Lowell's canals but,
just before Mariner 4 was launched, new spectroscopic data
were lowering previous estimates of the planet's atmospheric
pressure, and Mariner 4 indicated that the pressure was, in fact,
only *one two-hundredths* of that here; this meant that any liquid
water would instantly evaporate. In 1963, by telescopic means
from Earth, *some* water vapor had been detected in the Martian
atmosphere. But in 1969 Mariners 6 and 7 confirmed Mariner
4's data on atmospheric pressure, and Mariner 7 provided the
further news that the "ice caps" Lowell had relied on to irrigate
the planet contained almost no water at all, but were composed
of frozen carbon dioxide. The credibility of Lowellian ideas (as
the more flamboyant notions of life on Mars are still called) was
at low ebb indeed. The ideas were hard to give up, though. For
example, the Mariner 7 scientist who had discovered that the
Martian ice caps were made of carbon dioxide had initially
misinterpreted his data in a way that encouraged the concept of
planetary life. "He had made a very important discovery, but he
initially misread it, I feel, because of the expectation of seeing
something else," said Murray, who felt that imagination had
too long been an ingredient in the study of Mars. He declared at

the November 1971 symposium, during which he sometimes managed to sound very much like Alfred Russel Wallace:

> Mars somehow has extended and endured beyond the realm of science to so grab hold of man's emotions and thoughts that it has actually distorted scientific opinion. . . . The reason this has happened is that man . . . has been guilty of wishful thinking collectively. We *want* Mars to be like the Earth. . . . It has been very, very hard to face up to the facts, which have been emerging for some time, that indicate it really isn't that way, that it is just wishful thinking. It hasn't been just the science fiction writers who have used that deep-seated feeling among human beings to do other things and give other messages. The people who have really fallen on this have been the scientists themselves, who have misunderstood the significance of their observations. . . . My own personal view is that we are all so captive to Edgar Rice Burroughs and Lowell that the observations are going to have to beat us over the head and tell us the answer in spite of ourselves.

Geologists, who are the conservatives of space science, are the natural opponents of the more theoretical scientists like Sagan. They tend to consider the planets by extrapolating from their knowledge of rock formations on Earth—something Sagan, who criticizes the geologists as too Earthbound, condemns as argument by analogy—a technique Sagan feels is worthless in dealing with the unknown. Even more than most scientists, the geologists want to avoid considering anything that can't be proved, and because they cannot imagine that life as they know it on Earth could have developed under the conditions they see on Mars, they will have little truck with notions of extraterrestrial life until an example of it is laid in front of them; they dismiss exobiology as "the science with no subject matter." Sagan—in a reversal of Murray's argument—accuses the geologists and others of *not* wanting to find life on Mars, perhaps because it would throw into question certain geocentric notions of their own. The geologists, of course, think of themselves as realists. They have been carrying on a running battle with Sagan at least since the Apollo flights to the Moon, when Sagan used to argue that it was possible that there were microbes, if not on the lunar surface, then under it; he used to talk to the astronauts about how to find them. At one time, Sagan further suggested the possibility that lunar microbes might start an epidemic on Earth, and the geologists have always blamed him (unfairly, perhaps) for the quarantine of the material brought back from the Moon, which made their study of the rocks more difficult; when the lunar quarantine barrier proved leaky, Sagan accused the geologists of playing fast and loose with the biosphere. After Apollo 16, when men had been to the Moon five times and the geologists, and practically all the biologists and biochemists associated with the Apollo program, believed they had proved beyond all reasonable doubt that there could never have been life there, Sagan was still insisting that there might have been and might still be—that the astronauts and the geologists simply hadn't looked deep enough. The geologists have never forgiven him. The matter still rankled; when a few months before the launch of Viking 1 Sagan addressed a conference of lunar geologists—many of whom are also Martian geologists—in Houston, he committed the unforgiveable sin, in their eyes, of calling the Moon "dull." Several geologists thought this was a form of sour grapes. Sagan compounded his sins, in the eyes of the geologists, by saying that before they ever attempted to bring rocks and soil back from Mars—something they very much wanted to do, though Viking was not capable of it—NASA ought to demonstrate that there was no danger of exterminating life on Earth by first sending a capsule of anthrax germs, the most lethal known, streaking toward this planet at the speed of a capsule returning from space; land the capsule; and then run the germs through the quarantine system that would be used for Martian rocks. The geologists, who remembered the leakiness of the lunar quarantine, feared that Sagan was trying to exterminate *them*. Sagan, of course, retorted that the geologists were simply demonstrating that the problems of contamination were very real. "Carl really drives his colleagues up the wall!" one geologist said at the time. "Epidemics caused by bugs brought back from Mars is straight *Andromeda Strain* stuff. The statistical probabilities are unbelievably low. Christ, if you take a bug out of its environment, the problem is keeping it alive, not worrying about starting an epidemic!" Epidemics aside, feeling ran high on such matters not only because at stake were perceptions of entire planets—and beyond that, the possible population of the universe—but also because the dispute involved the way science is to be practiced on Earth.

When it came to Mars, the geologists, given the data gathered by the Mariner 4, 6, and 7 missions, might have been able to ignore Sagan had not Mariner 9, sent into orbit in 1971, revealed a rather different planet. Mariner 9 arrived in the middle of a Martian dust storm that hid nearly everything on the surface. The only features of much interest that Mariner 9 picked up were four black spots in the region of the equator. When the dust began to clear, the four black spots turned out to be the highest points on Mars, the cratered summits of four vast volcanoes—gigantic calderas—the largest of which was perhaps 18 miles high and far bigger than any mountain on Earth. They were clustered in an area of Mars called Tharsis, which the three earlier flybys had missed. Volcanoes have important implications for life on a planet because they are believed to be the source of most of the gases in the atmosphere (and therefore the water in the oceans); the more internally active a planet is, the more likely is the chemical activity that leads to life. Murray, who more than most physical scientists is concerned with individual motivation, wrote in *Mars and the Mind of Man*, after the Martian volcanoes were revealed:

> When the photographic evidence of these huge volcanoes first came in, I simply couldn't accept its significance. I too was a victim of the very process I described [earlier], of being so captured by the prejudices that had grown up in my own mind about the planet as to have great difficulty in accepting and understanding the significance of new data when it arrived.

And when the dust settled (which took about 2 months), the entire Tharsis region, constituting about an eighth of the planet, turned out to be a ridge, having built itself up by the outflow of magma from the calderas. Emanating from one edge of Tharsis, 2 to 3 miles deep and extending for more than 3,000 miles just south of the Martian equator, was a huge valley (later named Valles Marineris, after the craft that discovered it) that the earlier flybys had also missed; to some geologists it looked like the beginning of a system of rift valleys like the one that divides the plates of the Earth's crust, enabling them to drift about on the hot, liquid rocks below. Murray wondered briefly whether Mars might be a dead planet that was just coming to life. With respect to internal activity, Mars appeared a lot livelier than the Moon, though a good deal less lively than the Earth—a reasonable state of affairs, inasmuch as Mars is midway in size between the two. Like the Earth, Mars, it was now evident, had a number of distinct geographical regions, as would be the case with any complex planet, for in addition to the ancient cratered area that the earlier flybys had detected—

and to the volcanic Tharsis region and the great rift valley, which they had not—the Mariner photographs showed vast circular lowlands, suggestive of the *maria* on the Moon. Unlike the situation on the Moon, though, there were signs that water had at one time poured across the planet—giant streambeds that appeared to well from underground and vast floods, very likely from similar subterranean sources, had evidently rushed across wide areas, leaving washes, streaks, and even what appeared to be the remains of muddy islands and giant bars.

Given the apparent paucity of water on the surface of Mars today, the most striking news from a Lowellian point of view, when all the data were in, was that Mariner 9 had shown that both poles, while overlaid with frozen carbon dioxide for most of the Martian year (which is 687 days long) had permanent caps beneath, present throughout the summertime, that apparently contained some frozen water.

Mariner 9 had arrived when it was spring in the Martian south; the north pole was shrouded in the darkness of approaching winter, and the north polar ice cap lay beneath a gigantic cloud of carbon dioxide snow, called the polar hood, which did not dissipate entirely until the northern spring, in the last few weeks of the mission. (The seasons on Mars last longer than our own. Because of the pronounced ellipse of the Martian orbit, the seasonal durations of the two hemispheres are different; the northern winter, somewhat shorter than the southern winter, lasts 154 Mars days, or 158 terrestrial days.) However, the southern cap, which was expected to wane throughout the southern summer, retained in the next few months its 200-mile-wide outline; this is what indicated that its surface, at least, was made of frozen water, which does not evaporate as readily as frozen carbon dioxide. Meanwhile, just north of the equator, white clouds, similar to the ones Lowell and Schiaparelli had detected as seasonal white spots, began to form close to the largest of the calderas, Olympus Mons. Earlier astronomers, who knew nothing of the volcano, had chosen the name Nix Olympica, or Snows of Olympus, for the mysteriously appearing and disappearing white spot in its vicinity. The clouds were not caused by an eruption, but probably by condensation high in the atmosphere as warm air from below was forced up and over the caldera, just as clouds are formed over high mountains on Earth. Some Mariner 9 scientists thought that the clouds might be partly water vapor. Sagan, who was watching the photographs as they came in to see what had caused the seasonal variations Lowell had noticed, was fascinated. He said at the time: "There was no telling what might come in on the photos. It was a time of sustained excitement—like going down the Zambesi for the first time. Usually excitement lasts a short time, but this was a yearlong high! And Viking will be the same."

As a result of the giant global dust storm, light or dark streaks of dust emanated from the lee of the craters, creating what Sagan called "natural weathervanes and anemometers," since they made a weather map of the wind patterns on Mars. They fascinated Sagan as soon as he saw them. "Carl is always drawn toward the dynamic, and that's why he likes Mars," one of the geologists who had heard him disparage the Moon said. "Any day-to-day change, even if it's only windblown dust, interests him. On the Moon, there's nothing as lively even as dust storms." The Martian dust storm had been unlike anything known on Earth: for one thing, the atmosphere of Mars is so thin that it has to be moving quite fast—well over 100 miles an hour—to pick up a dust particle at all; for another, many of the particles were apparently so small that they stayed aloft for months after the wind stopped. Even in Lowell's day, there had been astronomers who said that the great seasonally changing splotches (not the white ones, but the blue-green ones, which he had claimed were planets) were not blue green at all but gray, and were caused not by vegetation but by windblown dust. Out of an attachment to the idea of Lowell's vegetation, though, a number of scientists tried to combine the two ideas, and as late as 1965 one microbiologist, Wolf V. Vishniac, whose mind worked in the same imaginative way as Sagan's, suggested that light-colored dust might settle on the dark leaves of plants after a sandstorm and subsequently slide off onto the ground, causing an area first to lighten and then to darken. In 1967, Sagan and James Pollack proposed an explanation for how the wind blows the dust about on Mars to make the seasonally changing patterns. Winds of varying velocities, they said, might pick up different-sized particles, sorting them, and different sizes might appear as different shades of gray; thus the various winds, which themselves are seasonal, would act as brushes continuously painting and repainting the planet. Observations made by Mariner 9 bore out their theory. Sagan noted, however, that there was "nothing in these observations that excludes biology," another of those (grammatically correct) uses of the double negative frequently employed by exobiologists. Sagan says he uses it in order to hold on to an idea in the face of the intolerance many of his colleagues show toward ambiguities in science. Sagan may also be fond of the double negative because it allows him to say things without really *saying* them—there is no doubt that it permits controversial statements to slip down more easily, in a tentative manner, so they can be retracted later.

Meanwhile, the geologists were beginning to read the history of Mars, using the Moon as a yardstick, for enough data had been gathered during the Apollo program to make the Moon a convenient point of reference for the solar system as a whole. There was some danger in this, for the geologists used the density of craters to judge the age of planetary terrains and the features on them, even though the meteoritic bombardment that caused the craters might have fluctuated from one part of the solar system to another. The Sun and the planets had evidently condensed from a huge gas-and-dust cloud, the solar nebula, about 4.6 billion years ago; the process of building up the planets, though, continued until about 4 billion years ago and culminated in giant impacts that left huge circular basins on them, such as the Mare Imbrium on the Moon, the Caloris basin on Mercury, and a basin called Hellas on Mars. The basins filled with lava welling up from the interior; on all three planets, radioactive material gradually heated the deeper zones, melting them; the heat moved gradually toward the surface. It was in this period, most geologists believe, that on Mars the cataclysmic flooding—the giant streambeds, the great washes— occurred, the water rising suddenly from a huge underground layer of permafrost or ice that had been melted geothermally— that is, by the newly generated heat inside the planet. In those days, Mars very likely had a thicker atmosphere—also the product of greater internal activity, volcanism, and outgassing— and this may have allowed the water to remain on the surface for a time as a liquid. Some of the water may have sunk back into the ground. On the Moon, the internal heat began to die down about 3 billion years ago, but on Mars, whose diameter is twice the Moon's, the heat continued far longer, building up the Tharsis ridge with the outflow from its giant calderas about 1.5 million years ago. Well before that time, though, the planet had cooled enough so it was not generating as much atmospheric gas. Some of the atmosphere leaked away into space. As Mars became increasingly glacial, atmospheric gases—mostly carbon dioxide, but water vapor as well— were frozen out at the poles, perhaps with some of the water vapor bound beneath the Martian surface as a layer of permafrost. Conceivably, the permafrost—which might also have contained the residue of the ancient subterranean ice—was planetwide (it has been described as a sphere of ice just inside

Mars) and came closest to the surface at the polar regions; the permanent, summertime polar caps might be aboveground manifestations of it. Most enticing to the Mariner scientists, there were thousands of meandering channels branching out from smaller and smaller tributaries, which suggested dried-up river systems of the sort that on Earth drain broad areas after rain. But they were of a sort distinctly different from the giant ancient streambeds. And there was nothing like them on the Moon. Although precise counts of the craters impinging on them or on the terrain they ran across had not yet been made to establish their age, they appeared to be much younger—indeed, some geologists thought some of them might date from relatively recent epochs.

All these signs of geological and meteorological activity were so tantalizing that the geologists were in favor of reducing the orbit of Mariner 9 in order to get a better look, but this suggestion, when it was made late in the mission, provoked an outburst from Sagan and another especially strong one from Joshua Lederberg. Both of them were now seeing Mars as a likelier habitat for life than ever and didn't want to endanger that habitat. "Lederberg turned ashen. He was afraid the spacecraft would crash if it got too close and would contaminate the planet with organisms from Earth," the geologist who had made the suggestion, Harold Masursky—the leader of Mariner 9's imaging team and a senior scientist at the Branch of Astrogeologic Studies of the United States Geological Survey, who is not a firm believer in life on Mars—recalled later. "It was very emotional." This was an understatement. Lederberg, along with Sagan, had sat on several international commissions that had dealt with the problem of keeping the planets uncontaminated by microbes from Earth. Before they were launched, the two Viking landers were baked for 80 hours, at a peak heat of

233° F, inside containers called *bioshields*, which were not removed until the two craft were safely on their way toward Mars. Mariner 9, which was never intended to land on the planet, was not sterilized, however.

Lederberg, who was then chairman of the Department of Genetics of the Stanford Medical School and was a member of Viking's biology team, had won a Nobel Prize in 1958—he shared the prize in medicine and physiology that year with two other American geneticists—for the discovery that bacteria reproduce sexually; he is therefore one of the men who over the last 25 years or so has shown the steps by which more complicated life forms on Earth (and so, perhaps, elsewhere) developed. Before the current tide of biological and biochemical discoveries, many people had assumed that organic material on Earth had always existed and consequently that questions about its origin were pointless; they did not seriously consider at all the matter of whether life might exist anywhere else. Lederberg had become interested in the matter when, as an undergraduate at Columbia in the early 1940s, he read a book called *The Origin of Life on Earth*, published in 1936 by a Soviet biochemist named Alexander Ivanovich Oparin, who hypothesized that organic compounds derived from atmospheric gases under the influence of sunlight. "Oparin opened up for me the idea of life on other planets by indicating that once the development of life is shown to be a chemical process it can happen anywhere that the conditions are right," Lederberg has said. Lederberg took Oparin's ideas a step further and, incidentally, influenced Sagan's thinking. Because Mars, among the extraterrestrial planets men can reach, was the most likely to harbor life and so offer the best, if not the only, chance of directly testing the molecular biologists' new theories, Lederberg was unalterably opposed to risking the impact of an unsterilized spacecraft. Sagan backed him by conjuring up visions of terrestrial organisms from the wreckage being blown by the winds all over Mars, possibly obliterating Martian biology and, furthermore, quite capable of deceiving future missions to the surface about the presence of indigenous life.

Sagan and Lederberg, in their search for life, were encouraged by the Mariner 9 photographs showing the many meandering tributary channels; it was difficult for them, and most others, to conceive of any way the channels could have been made except by the drainage of broad areas after rain. Most biologists agree that water is a concomitant to life on a planet because organic compounds require a solvent in order to react with each other and with inorganic compounds, and water is by far the most cosmically abundant (and efficient) solvent. Extraterrestrial life will most likely be organic, they agree, because carbon is far and away the most versatile element in its ability to bond with other elements to form the complex compounds living things are made of—particularly the long polymer chains essential for the encoding of genetic information. The Martian tributary-type streambeds seemed to be concentrated near the equator, where even in the present era the temperature is warm enough so that during part of the day water could exist in liquid rather than frozen form. The temperature of the Martian equator at noon may reach as high as 90° F. There is, of course, too little atmospheric pressure for liquid water to exist on the surface today, but Sagan felt that Mars even in relatively recent epochs may have possessed a heavier atmosphere in which water would not have evaporated. There was evidence that over long periods of time conditions on Mars had changed, for around the polar caps there was a broad series of curious laminations, as though Mars were a gigantic onion partly peeled at both ends; one theory about how the laminations were made is that in the course of successive ice ages, the caps were covered with dust, and as they retreated the dust was deposited in concentric rings. Whether or not the laminations were made from deposition, they did appear to be evidence that the Martian climate is capable of broad swings. On the basis of celestial mechanical calculations, scientists had concluded that the orientation of Mars's axis slowly alters over a period of at least 50,000 years, so that first one and then the other pole is more directly exposed to sunlight. Sagan had proposed before the Mariner 9 mission that the caps would therefore melt alternately, releasing carbon dioxide into the atmosphere and creating enough pressure so that whatever water there was on Mars could be in a liquid state, and also making the planet warmer. This was not entirely a new idea, for in 1941 a Yugoslavian astronomer, M. Milankovitch, had published his theory that a similar movement of the Earth's axis, known as the *precessional cycle*, was responsible for the successive ice ages here. Nevertheless, Sagan was the first to apply the theory to Mars. He postulated that the Martian carbon dioxide would remain in the atmosphere for about 12,000 years, or a quarter of the cycle, until the other pole had cooled sufficiently to freeze it into a new cap; there the carbon dioxide would remain for 12,000 more years, during which Mars would be dry and desolate—as it is now—and then would warm that pole in its turn, releasing the gases into the atmosphere to cause another 12,000-year interglacial period. In one of the more Lowellian passages of *The Cosmic Connection*, Sagan wrote:

Twelve thousand years ago may have been a time on Mars of balmy temperatures, soft nights, and the trickle of liquid water down innumerable streams and rivulets, rushing out to join mighty, gushing rivers. Some of these rivers would have flowed into the great [Marineris] rift valley.

If so, twelve thousand years ago was a good time on Mars for life similar to the terrestrial sort. If I were an organism on Mars, I might gear my activities to the precessional summers and close up shop in the precessional winters—as many organisms do on earth for our much shorter annual winters. I would make spores; I would go into cryptobiotic repose; I would hibernate until the long winter had subsided.

This was too much for Murray. In an essay in *Mars and the Mind of Man,* written after Mariner 9's revelations, he had conceded handsomely that "Mars turned out to be different from what any of us thought, and demonstrated that I personally have, to some extent, been a victim of my own prejudices about the planet." In the next sentence, though, he added, "I still find myself on the less optimistic side about the possibility of life on Mars." At first, he had thought the calderas were evidence that Mars was just beginning to become active; later, when evidence turned up of earlier calderas that had been largely eroded, Murray decided that Martian volcanism has been going on for perhaps two-thirds of Mars's 4.5-billion-year history, the view currently held by most geologists. "Even the mysterious channels seem to me to represent a brief episode in the history of the planet," he said. Sagan's rhapsody on "streams and rivulets, rushing out to join mighty, gushing rivers" led him to ponder the difference between Sagan and himself:

As each of us draws opposite conclusions about the significance of the Mariner data regarding life on Mars, it is difficult to escape the impression that we are both interpreting it in the context of differing a priori views of the planet.

Since Sagan and I respect each other greatly as scientists and find much stimulation in each other's thoughts, why should we find it so difficult to read the record similarly? One can look first to our scientific backgrounds. He aimed at the planets and research from his undergraduate days. My first love was—and is—the Earth, and my initial postgraduate activities were of an applied nature. I didn't return to a university for a research career until I was twenty-nine years old. Carl had emphasized synthesis and conjecture about how things are, might be, or could be beyond the Earth. If he is lucky, his great passion, the search for extraterrestrial life, especially intelligent life, will blossom during his lifetime. I, on the other hand, have been mainly concerned about distinguishing fact from fiction in a subject mouldy with misconceptions and inherited prejudices. My passion is to understand how things *really* are on Earth as well as in space.

Both Sagan and Murray were engaged in what has become a more and more common way of conducting scientific inquiry—a process known as *model building*, in which hypothetical models are constructed until one turns up that fits the problem at hand. It is a process people used to call simply "thinking" or "having ideas." The difference, if there is one, lies in the assumption that the mind operates like a computer, sorting rapidly through a number of alternatives until it finds the right one; in fact, *model building* is part of computer terminology. If the scientific computer is to proceed efficiently—or so the theory goes—it must have the widest variety of models to choose from. But the very use of the term has led to another difference: since most of the models in question are made in order to be discarded for better ones, the builder allows himself to be highly imaginative in the course of their construction; compared with ideas, models are easy to give up, or at least they are supposed to be. Though flexibility of thought and the capacity to keep more than one possibility in mind are essential in science, thinking of ideas as models has its dangers, for the most far-out notion gains a sort of acceptability by virtue of the fact that the man who devised it can say he is engaged in model building.

In the years immediately after Mariner 9, many scientists, both geologists and biologists, began chipping away at Sagan's model of Mars, with its warm, watery interglacials and its possibility of sporelike life. Harold P. Klein, the director of life sciences at NASA's Ames Research Center and the leader of Viking's biology team, said of Sagan's interglacials: "If that's the only way to get water on the surface of Mars, I'd say it further reduces the chances of finding life there."

Thomas A. Mutch, then a geologist at Brown University and the leader of the lander imaging team (he is now NASA's associate administrator for space science), expressed caution about the underpinnings of the model—Milankovitch's theory, which had, in fact, never been proved to be the explanation of our own glacial periods. Although his theory is now favored, others come forward from time to time, such as that the glacial epochs are the result of bursts of volcanism or changes in the Sun's brightness. "The point is, if we can't solve this matter for the Earth, then you have to be skeptical of this model for Mars," said Mutch.

The permanent polar caps, which were supposed to contain the gases Sagan needed for his warm, watery interglacials, covered a disappointingly small area. Sagan, however, believed they were several kilometers deep and contained enough carbon dioxide to create, if it should be released as a gas, an atmospheric pressure equivalent to the Earth's, or about 200 times the present Martian pressure. Murray, on the other hand, claimed that the permanent caps were so very thin that if their gases were released the pressure of the Martian atmosphere would be increased only 5 or 6 times. One of the sticking points of the dispute was the appearance of the summertime cap. From overhead, in the Mariner 9 photographs, the cap looked like a patch of slushy snow with curving paths running through it of the same color as the surrounding sands of Mars. Murray claimed that the paths were the bare ground showing through the cap, which, therefore, had to be very thin; indeed, the configuration of the paths matched the terrain around the cap. Sagan said that what looked like bare ground was not bare ground but an amalgam of windblown dust and ice on top of a thick cap and that the reason it resembled the surrounding terrain was that both were laid down by the wind patterns swirling around the poles. The prevailing winds on Mars do in fact remove the dust from the equator and deposit it in the polar regions more deeply than anywhere else. This situation led Sagan to speculate about another way in which the caps could be melted. If enough dust landed on a cap, Sagan said, it would darken the cap, so that the sunlight would warm it, eventually melting the ice and releasing its contents to the atmosphere.

Unlike Lowell, who was inflexible in his ideas, Sagan was always ready to move on to other models—even if it meant adapting one of his opponent's. He said that if Murray's model should happen to be correct, and the permanent caps turned out to be very thin, they would still supply enough atmosphere to provide the pressure to keep liquid water on the planet. Murray, however, was not prepared to make Sagan a present of a Mars with "balmy temperatures, soft nights, and the trickle of liquid water down innumerable streams and rivulets." He didn't believe the channels were cut by recurring periods of rain within anything like the time period Sagan invoked, since there was, in his opinion, no sign that rain had fallen on Mars in recent ages. Geologists had been busy for some time doing crater counts of the tributary streams, a task that presented some difficulty as it was hard to establish how many craters actually impinged on the streams. However, the evidence now seemed to indicate that many of these channels, far from being recent, were in fact quite old: in some cases they were very heavily cratered indeed. Furthermore, there were no signs of stream-beds on Olympus Mons, which appeared to be about 0.5 billion years old, and Murray took their absence as proof that it has not rained since that time. Sagan, however, said there were streambeds on another big volcano just north of the Tharsis region, Alba Patera—but Alba Patera was thought to be older than Olympus Mons and was also an atypical feature still puzzling to the geologists. Some geologists came down in the middle, saying that although the channels might have been carved by successive wet periods, as Sagan suggested, these were separated not by a few thousand years but by perhaps millions of years and that the last wet period was perhaps hundreds of millions of years ago. Sagan, while not accepting their hypothesis entirely, did admit that if he were a Martian spore hibernating in cryptobiotic repose, as he had once imagined, he would have trouble waking up after all that time.

Unfortunately for Sagan's theories, no one came up with an age for the Martian streambeds of less than 100 million years (a figure suggested by a young colleague of his at Cornell), obliging him to recognize that the news from the geologists made recurrent Earthlike conditions on Mars, tied to the precessional cycle, not quite so likely. In an effort to save his model, Sagan later dropped the precessional cycle as an explanation of recurrent warm, wet periods on Mars and substituted another explanation—one that he had had on the back burners for several years. He suggested that a bobbing movement of the axis of Mars's rotation (an effect of the gravitational tugs of Jupiter) caused alternate glacials and interglacials on Mars of about 0.25 million years—substantially longer than the few thousand years he originally proposed, but not nearly long enough to satisfy even the most optimistic geologists' evidence. "That doesn't say you can't have a trickle of water from time to time—like the Colorado River down the Grand Canyon," he said. Should it turn out that there has been no Colorado River or anything like it on Mars in the last few hundred million years, Sagan was prepared to retreat to drier, if not higher, ground. "I think life can do quite well without water. In fact, the deserts of Mars are really *oceans*—that is, if the wee beasties there can tap the water in the rock. The chemical bond that unites water to rocks is very weak; one or two electron volts is all that you need to break it, and a single near-ultraviolet photon of light, such as falls regularly on Mars, has that same amount of energy. So a Martian organism might use sunlight to snip the water off the rock. On Earth, there are kangaroo rats who never drink because they can derive their water from their food chemically; though they don't get their water from rocks, the idea for Mars is not very different. So even if there is no evidence for flowing water on Mars, I'm still sanguine about life." In presenting his argument, Sagan had the look of a man who has made a last-ditch effort and knows it; he added: "You understand the spirit in which I'm saying these things? They're within the bounds of permissible speculation, but nothing I insist on." Some of his Viking colleagues, however, thought he took his models a little more seriously than these words suggest. "Sagan struggles to create situations where life might exist," one geologist said. "It's a compulsion."

Sagan was born in 1934 in the Bensonhurst section of Brooklyn, where his father was a cutter in a clothing factory. "It was during the Depression, and we were kind of poor. When I was very little, the basic thing for me was stars. When I was five years old, I could see them at whatever time bedtime was in winter, and they just didn't seem to belong in Brooklyn. The Sun and the Moon seemed perfectly right for Brooklyn, but the stars were different. I had the sense of something interesting, distant, strange about them. I asked people what the stars were, and I mostly got answers like, 'They're lights in the sky, kid.' I could tell they were lights in the sky; that wasn't what I meant. After I got my first library card, I made a big expedition to the public library branch on Eighty-sixth Street in Brooklyn. I had to take the streetcar; it was some big distance. I wanted a book on the stars. At first, there was some confusion; the librarian mentioned all kinds of books about Hollywood stars. I was embarrassed, so I didn't explain right away, but finally I got across what I wanted. They got me this book, and I read it right there, because I wanted the answer." (Sagan was a precocious child. So is his son Nicholas, who resembles Sagan in many ways. Nicholas taught himself to read by the age of 21 months—a fact his parents discovered when he began rattling off the road signs on a transcontinental car trip. Sagan is an indulgent father, and on occasion, he can be an ingenious one; for example, on that trip, during which he dictated huge sections of *The Cosmic Connection* into a tape recorder, he also played back a number of children's stories he had taped earlier, which kept the boy occupied between road signs. Nicholas is indeed wise beyond his years; just before the launch of Viking 1, when he was asked whether he believed there was life on Mars, he replied, "Maybe yes and maybe no.") Sagan continued: "The library book had this stunning, astonishing thing in it—that the stars were suns, just like our Sun, so far away that they were only a twinkle of light. I didn't know how far away that was, because I didn't know mathematics, but I could tell only by thinking of how bright the Sun is in the daytime and how dim a star is at night that the Sun would have to be very far away to be just a twinkle, and the scale of the universe opened up for me.

"It must have been a year or two after this that I learned what the planets were. Then it seemed absolutely certain to me that if the stars were like the Sun there must be planets around them. And they must have life on them. This was an old idea, of course. Christiaan Huygens, the Dutch astronomer, I found out later, had written about it in the 1670s. But I thought of it before I was eight. And once I reached that point, I got very interested in astronomy. I spent a lot of time working on distances, coordinates, and parallaxes.

"Then, when I was ten—I was at P.S. 101 in Brooklyn at the time—I came upon the Edgar Rice Burroughs novels about John Carter and his travels on Lowell's Mars. It was a world of ruined cities, planet-girdling canals, immense pumping stations—a feudal technological society. The people there were red, green, black, yellow, or white, and some of them had removable heads, but basically they were *human*. I didn't realize then the chauvinism of making people on another planet like us; I simply devoured what seemed to me the riches of another planet's biology. Carter fell in love with a princess of the Kingdom of Helium, Dejah Thoris. It was very exciting, and I loved those books. They were full of new ideas. On Burroughs's Mars, there were two primary colors more than on Earth, and I would close my eyes and try to imagine them. I tried to imagine my way to Mars, the way Carter did: I would go into a vacant lot, spread my arms, and wish to be on Mars."

A few months before Viking was launched, Sagan taped up on the wall outside his Cornell office a map of Mars as Burroughs portrayed it, with Xs marking the spots where Carter landed. In his office, he showed a visitor, on a globe of Mars made from Mariner 9 photographs, exactly where Carter would have come down. "Many an evening I spent in vacant lots, arms outstretched, *thinking* myself to that twinkling red place, but nothing happened. I tried all different kinds of wishing. Suddenly, it dawned on me that this was fiction; maybe there was some better way to get to Mars.

"This was toward the end of the Second World War, and I heard about the V-2 rockets the Germans used to bomb England. There were occasional references in the papers to what rockets could do for space technology. I found some journals put out by a group called the British Interplanetary Society. It sounded nice. And gradually I realized that there *was* a way: I discovered that in 1939 the British Interplanetary Society had published a study for a multistage rocket that could go to the Moon. If the Moon, then why not Mars?

"I didn't make a decision to pursue astronomy; rather, it just grabbed me, and I had no thought of escaping. But I didn't know that you could get paid for it. I thought I'd have to have some job I was temperamentally unsuited to, like door-to-door salesman, and then on weekends or at night I could do astronomy. That's the way it was done in the fiction I read, in which space science was practiced by wealthy amateurs. Then, in my sophomore year in high school, my biology teacher (this was at Rahway High, because we had moved to New Jersey) told me he was pretty sure that Harvard paid Harlow Shapley a salary. That was a splendid day—when I began to suspect that if I tried hard I could do astronomy full time, not just part time.

"I had been receiving catalogues from various colleges, and I wanted one with good mathematics and physics. The University of Chicago sent me a booklet entitled *If You Want an Education.* Inside was a picture of football players fighting on a field and under it the caption, 'If you want a school with good football, don't come to the University of Chicago.' Then there was a picture of some drunken kids and the caption: 'If you want a school with a good fraternity life, don't come to the University of Chicago.' It sounded like the place for me. The trouble was that it had no engineering school, and I wanted an education not only in astronomy and physics but also in rocket engineering. I went down to Princeton to ask Lyman Spitzer, the astronomer, his advice; he was involved in some early rocket studies. He told me that there was no reason an astronomer had to know every nut and bolt of a spacecraft in order to use it. Up until then, I had thought this was necessary—another holdover from the fiction I'd been reading, in which the rich amateur built his own spaceship. Now I realized that I could go to the University of Chicago, even though it had no engineering school. I applied and entered in the fall of 1951. In the early 1950s the University of Chicago was a very exciting place to be. It was strong in the humanities—which I wanted—but it was also very strong in the sciences. Enrico Fermi and Harold Urey were both there, in physics and in chemistry. And it had a superb astronomy department, which operated the Yerkes Observatory."

Sagan began to attract the attention of older scientists, many of them Nobel laureates, who, after experiences in a variety of fields, were beginning to think about extraterrestrial life. Back in Rahway for Christmas vacation during his freshman year, he met a young biologist—the nephew of a friend of his mother's—who was at Indiana University, working with H. J. Muller, who had won the 1946 Nobel Prize in Medicine and Physiology for the discovery that X-rays caused mutations in genes; Sagan was interested because X-rays are produced by exploding stars—novas or supernovas—and Muller's discovery showed them to be a direct link between astronomy and the evolution of life. The young biologist told Sagan that Muller was now working full time on the origins of life. Later, back at the University of Chicago, Sagan wrote his new friend a letter; the friend showed it to Muller, who liked it and wrote to Sagan asking him to spend the summer of his freshman year working for him at Indiana. "Muller had me doing routine things, such as looking at fruit flies for new mutations," Sagan said. "But he ran a real research group, and for the first time I got a feeling of what scientific research was like. Moreover, Muller was interested not only in the origins of life but in the possibility of life elsewhere; he didn't think the idea was the least bit silly."

Muller, of course, was by no means the first biologist to concern himself with extraterrestrial life; that distinction proba-bly belongs to Alfred Russel Wallace (who, although he

attacked the notion with respect to Mars, at least took the
overall proposition seriously)—and the tradition continued
more affirmatively with J. B. S. Haldane and Alexander
Oparin. "Muller encouraged me to learn genetics," Sagan went
on. "Later, he sustained me through years of studying biology
and chemistry, which I had thought were far removed from my
main interest, astronomy. I always kept in touch with him. A
few years before his death, he gave me a book about space flight
by Arthur C. Clarke, and inscribed it, 'Perhaps we'll meet
someday on the tundras of Mars.' He died in 1967. In 1973,
after Mariner 9 had mapped Mars, I managed to get a crater
there named for him." Sagan is a member of the Subcommittee
on Martian Nomenclature of the International Astronomical
Union, which now handles such matters; he also participated in
naming craters after Lowell, Schiaparelli, and Edgar Rice
Burroughs.

   In the fall of his sophomore year, Sagan returned to the
University of Chicago with a letter from Muller introducing
him to Urey, who had won the 1934 Nobel Prize in Chemistry
for the discovery of heavy hydrogen and had gone on to the
study of the origins of life. (This was the basis of Urey's interest
in lunar science, a field of which he is generally considered the
modern-day founder.) Urey is a mild-mannered man with a
benign intelligence. "He was extremely kind to me when I was
an undergraduate," Sagan said. "I did an honors essay on how
life began. It was very naive, and I remember Urey's comment:
'This is the work of a very young man.' I had the idea that in
one fell swoop I could understand the origins of life, though I
had not had much chemistry or biology. It was an attempt to
learn by doing. Some other people at Chicago were more
effective at this time than I was. It was a time of great
excitement, for this was when Stanley Miller was doing his
work, under Urey, on the origins of life. He had filled a flask
with methane, ammonia, water, and hydrogen—things you

would expect to find in the primitive atmosphere of a young planet—and then had passed an electrical discharge, like lightning, through it. The result was amino acids, the first step toward life. Miller had shown that the beginnings of life were not a matter of chance, but could happen in any place where conditions were right." (A while ago, Sagan showed a visitor to his laboratory at Cornell a version of Miller's experiment, with further ramifications; the resulting organic compounds were a reddish-brown gunk that gave the same spectrum as the reddish-brown belts on Jupiter.) He went on: "Urey showed me through Miller's laboratory. Later, Miller was forced to defend his work before the University of Chicago's chemistry department. They didn't take it very seriously; they kept suggesting that he had been sloppy, leaving amino acids all over his laboratory. I was outraged that something as important as that could be received in such a hostile way. Urey was the only one who spoke up for him. He said: 'If God didn't create life this way, He certainly missed a good bet.'" Urey, who was 83 when Viking 1 was launched, was for a time a member of the team that would be looking for organic compounds on Mars.

After getting his master's degree in physics, Sagan went on in 1956 to the University of Chicago's graduate school in astronomy, which was at Williams Bay, Wisconsin. There he worked with Gerard Kuiper, a Dutch astronomer who at the time was the only full-time planetologist in the United States. Though Lowell had alienated the astronomical community, he wasn't solely responsible for planetary studies having fallen into disrepute. With the advent of astrophysics, in the 1920s, astronomy had taken a more professional turn, one that led away from the planets; indeed, Kuiper, whom Sagan saw as providing a link between the earlier planetary astronomers and what he called "the present burgeoning time," had to start by studying the stars before he could go into planetary work. In 1944, Kuiper had discovered that Titan, the largest moon of

Saturn, had an atmosphere, and that the major constituent of Titan's atmosphere was methane—a discovery that impressed Sagan, who later suggested that life might be found there. Kuiper was responsible for the idea that there might be lichens on Mars because he had found that the planet's spectrum was not inconsistent with them (though it was inconsistent with green plants). Sagan didn't think much of the idea because he didn't believe that specific terrestrial organisms would be duplicated in the course of Martian evolution. Nevertheless, he welcomed the suggestion, if only for its propaganda value, because he was already becoming interested in making the theory of extraterrestrial life acceptable again. "Kuiper was a respected man, and if he said it was possible for *any* sort of life to exist on Mars that was important," Sagan said. "It was a tremendous boost to exobiology." Sagan had spent the summer of 1956 with Kuiper at the McDonald Observatory, in Fort Davis, Texas, and there he had his first opportunity to see what Mars looked like in a close opposition through a big telescope. "As it turned out, there were dust storms in both places—Mars and Texas," he said. "I didn't find any canals. I was satisfied just to be able to see light and dark markings. The seeing was poor, even through the eighty-two-inch telescope at McDonald. There Mars was, though shimmering, squashed, distorted. Then, for an instant, the atmosphere steadied, and I caught a glimpse of the southern polar cap. I saw no fine details. It was no big deal. I realized that the telescopic technique, while interesting, was limited: sitting under a blanket of air forty million miles from the target was not going to tell me much."

While Sagan was getting his doctorate under Kuiper, he married a young biologist (they had two sons, Dorion and Jeremy, both of whom at the time of the Viking mission in the mid-1970s were in high school), and the couple moved from

Williams Bay to Madison, Wisconsin, whose university atmosphere was more to his liking. It was during this period that Sagan met Lederberg, then professor of genetics at the University of Wisconsin. Lederberg, who had just won his Nobel Prize, had a reputation for brilliance and inaccessibility. "He was an object of some consternation and fear," Sagan said. "Postdoctorates in biology were afraid to present papers lest he be in the audience and demolish their thesis with two questions. Then, one day, he called me up out of the blue and said he wanted to see me—said he was interested in extraterrestrial life. I was immensely flattered." One of the ideas Lederberg wanted to talk about was a thought he had had some years before, when he first read Oparin's work *The Origin of Life on Earth*. It had occurred to Lederberg that since 90 percent of the universe is made up of the same few atoms that are required for life on Earth—the main ones being hydrogen, carbon, nitrogen, and oxygen—then there should be no discontinuity in the development of organic compounds in the course of the evolution of the universe; that is, the compounds should be as common anywhere else—even in the space between the stars—as they are here. His theory was subsequently borne out: in the next decade, radio astronomers found clouds in interstellar space containing 40 or 50 varieties of organic compounds, such as formaldehyde, and it was discovered that some meteorites contain them too. Lederberg wondered whether postulating the generation of amino acids on each planet by means of lightning in its atmosphere, in the manner that Stanley Miller had demonstrated, was necessary; they could be raining down on the planets all the time. "What I like about working with Lederberg is that things turn out well both for biology and for astronomy," Sagan has said. The two scientists immediately took to each other. Sagan has described Lederberg as "a dry, stimulating, totally unfettered man who is willing to carry his ideas to their logical consequence, even though the prevailing wisdom says it's silly" and as "a natural resource that should be used widely." He soon regarded Lederberg as a mentor and a collaborator. "We stimulated each other's ideas," he said. "It was a pleasure to talk to him; we didn't either of us have to finish sentences. We could leapfrog through arguments—an efficient way of talking. Then and since, I've had dozens of ideas that I've been able to bounce off him. Many are jointly arrived at, so that neither of us knows which thought of them first."

Lederberg concurred in much of this. "Sagan has fired up some of my ideas, and I think I've helped fire up some of his," he said. "Back in the early days of our friendship, I think it was helpful to Sagan that an established biologist could entertain the same thoughts as he did about extraterrestrial life."

In 1959, Lederberg headed a committee of the Space Science Board of the National Academy of Sciences to study ways of searching for life in space; he asked Sagan, who was then 25, to be a member. The group included several men who would be important to Sagan later, including Wolf Vishniac, who was then at Yale and who shared many of Sagan's and Lederberg's ideas. Sagan, Lederberg, and Vishniac participated in many other conferences on extraterrestrial life, most notably a symposium sponsored by the Space Science Board in 1964–65, whose proceedings were published in a book, *Biology and the Exploration of Mars*, which provided an important scientific underpinning for Viking.

Sagan received his doctorate in astronomy and astrophysics from the University of Chicago in 1960; he went on to become a research fellow at the University of California at Berkeley, and then, at Lederberg's invitation, he spent a year as visiting assistant professor of genetics at the Stanford Medical School, to which Lederberg had recently moved. From 1962 until 1968, Sagan held a joint appointment as astrophysicist at the Smithsonian Astrophysical Observatory, in Cambridge, Massachusetts, and lecturer and later assistant professor of astronomy at Harvard. In 1968, he moved to Cornell, where he is currently not only the David Duncan Professor of Astronomy and Space Sciences but also director of the Laboratory for Planetary Studies. In addition to the commissions and conferences on extraterrestrial life, Sagan has participated in the work of an enormous number of boards and committees having to do with space exploration, including the groups that formulated the international procedures for sterilizing spacecraft and several committees for NASA—most notably, of course, the imaging teams of Mariner 9 and Viking.

His scientific achievements include over 200 papers, some of them written in collaboration with Lederberg or with Cornell associates. Some of the more imaginative ones concern such matters as how microbes might exist below the lunar surface; how life might exist in the clouds of Venus; how life might exist in isolated areas on Mars; how Martian microorganisms might survive the rigors of the planet by existing a centimeter below the surface; how the climate of Venus, enveloped in a hot, heavy atmosphere of carbon dioxide, which acts like the glass of a greenhouse, could be improved by dropping a certain type of algae into the clouds, which would break down the carbon dioxide and cool the planet; and how the Martian climate (if it was determined that there was no indigenous life) could be improved by depositing colonies of dark-colored microbes on the ice caps which could multiply and melt them (if Martian life was found, so that one would not want to introduce terrestrial microbes, carbon black could be substituted, and he had even calculated the number of rocket ships that would be required to get it there in sufficient quantities). A large number of Sagan's scientific papers up to that time had to do with communicating by radio with intelligent life elsewhere in the cosmos, and currently he was investigating this possibility with Frank Drake. Drake was director of the radio telescope at Arecibo, Puerto Rico, the biggest in the world, which was operated by Cornell for the National Science Foundation; the two scientists listened, when they had the time, for signals from intelligent life in outer space. "Sagan desperately wants to find life someplace, anyplace—on Mars, on Titan, in the solar system or outside it," one of his Viking colleagues said. "In all the divergent things he does, that is the unifying thread. I don't know why, but if you read his papers or listen to his speeches, even though they are on a wide variety of seemingly unrelated topics, there is always the question, 'Is this or that phenomenon related to life?' People say, 'What a varied career he has had,' but everything he has done has had this one underlying purpose."

Sagan was asked once why he thought it was that he, and others, are so interested in trying to find life beyond the Earth. "I think it's because human beings love to be alive, and we have an emotional resonance with something else alive, rather than with a molybdenum atom," he said. "Why are people interested in the life history of the armadillo? Why do we go to Antarctica to find out what the emperor penguins have been doing lately? It's fun, because we are primarily drawn to things that are alive."

Squamous Purple Ovoids

As is customary with NASA, so much planning went into the landings on Mars that the anticipation was almost more real than the event itself. The decisions to be made involved fundamental disputes over the nature of Mars as well as between the scientists themselves; the actual landing, when it occurred, would be almost tame in comparison. If all went according to plan, the lander section of Viking 1 would separate from the craft's orbiter section (the orbiter would continue to circle the planet) and head down to the surface of Mars—a trip that would take about 3.5 hours. When it reached an altitude of about 800,000 feet, the lander, inside a sort of pod called the *aeroshell*, would enter the Martian atmosphere, traveling at a speed of more than 10,000 miles an hour and planing with an amount of lift that would keep it from plunging too steeply and overheating. (That other, outermost pod, the bioshield, would have come off long ago, of course—soon after the spacecraft was beyond the Earth's atmosphere and hence out of danger of contamination by terrestrial organisms; so effectively did the bioshield seal the spacecraft that were it left on the ground controllers would have been unable to communicate with Viking on the 11-month voyage to Mars.) It was expected that the lander, still inside its aerodynamically shaped aeroshell, would plane for well over 300 miles across the planet, getting ever closer to the ground; at about 20,000 feet, when the speed was down to 560 miles an hour, the aeroshell would pop off and the craft's parachute would unfurl. The landing site had to be at a low enough altitude so that Mars's rarefied atmosphere would be sufficiently dense to slow the craft by means of the parachute; this meant the elimination of about three-quarters of the planet from consideration. Inevitably, some promising sites had to be ignored, but even so the scientists working with NASA were far from unhappy because the low sites were the ones most likely to have moisture and thus perhaps life. Years ago, Sagan and Lederberg had suggested that on Mars, which is generally far drier and colder than any spot on Earth, there might be *microenvironments* or isolated spots that would be particularly hospitable to life; presumably, they would be low areas, such as the bottom of a chasm, where there might once have been a river or where the warmth of the rocks might have melted underground permafrost geothermally to make a subsurface pool.

Since it is hard to land a spacecraft in a chasm, a safer site was picked for Viking 1, which, floating on its chutes, would (if all went as expected) head for a flat basin or low-lying plain in the northern hemisphere known as Chryse, to a spot that looked as if it was just beyond the mouths of four dried-up streambeds—not the tributary variety, but the giant ancient sort that date from the age of planetary inundation. One of the streambeds, Simud, appeared to originate in a chasm, which in turn led from the Valles Marineris, the great rift valley of Mars. The three others, however, had their sources in chaotic, broken-up terrain, which many geologists thought might be where in ancient times water had welled up when permafrost was melted geothermally. With the permafrost depleted, the ground would eventually have collapsed, to form the chaotic terrain there now.

In the spring of 1976, as the day of the landing grew nearer, many scientists, Sagan among them, began hedging their bets about water having been what carved out the Martian channels; at a scientific conference at the Kennedy Space Center at Cape Canaveral almost a year earlier, just prior to the launch of Viking 1, Sagan himself had pointed out that there were at least four other possible explanations for the channels' origin: excavations by wind-transported sand, cracks in the ground, liquid carbon dioxide, and flowing lava. Nevertheless, Sagan and most other scientists still considered water the most likely agent.

Vikings 1 and 2 would be landing when it was summer in the north of Mars, the best season to look for water and life. The lander of Viking 2, which would touch down about 2 months after Viking 1, was slated to land 1,000 miles northeast of Chryse, on a vast lowland called Cydonia, which was selected because it was in a latitude that was more likely than anywhere else in the north to have water at this time of year. There were alternative sites for both landers; the final decision about where each would touch down would be made a few days before landing. According to the latest estimates, the rarefied Martian atmosphere was ordinarily one one-thousandth water vapor—a figure that seemed to rise to one two- or three-hundredth in summer around noon in midlatitudes. Crofton Farmer, head of Viking's water vapor mapping team, felt that the best explanation for the increment in vapor was that on top of the layer of permafrost which increasingly many scientists now thought must be underground, unchanging, there might be a second seasonal layer of frost—he called it *tempafrost*—which vaporized in summer at midday and re-formed as the temperature dropped.

The cameras aboard Viking 1 would not yet be working as the craft descended toward Chryse, but if they could be, the visible differences between the surface of Mars and that of the Moon would be very apparent: fewer small craters would be seen on the Martian ground below than were seen on an approach to the Moon, which is virtually airless; the Martian atmosphere burns up small meteors, letting only the larger ones hit the surface. Moreover, the wind tends to smooth over the craters, further reducing their number. On an approach to the Moon, the number of craters appears to increase all the way to the ground, but on an approach to Mars the number would appear to stop increasing at a certain altitude, so that the general aspect of the terrain would remain the same the rest of the way down. Streaks of dust that the wind had deposited in the lee of the larger craters would provide another way for a space traveler to know he was approaching Mars, not the Moon. The ground is smoother than a lunar *mare*, and since it is reddish orange, it would look more like a terrestrial desert. Finally, Sagan suggested that Mars might be less like the Moon and more like the Earth in a most important way: life might exist there, even if it weren't evident to an approaching spacecraft. After all, if the spacecraft were landing in the Sahara, another place that appears quite barren even from low altitudes, the desert would upon closer inspection turn out to contain date palms, camels, Arabs, and oases—microenvironments of a sort. In *Icarus*, the scientific journal Sagan edits, he and a colleague published a paper in 1971 saying that the three earlier American missions to Mars, during which Mariners 4, 6, and 7 flew past and photographed the planet, afforded no chance at all of discovering even a civilization as advanced as our own, and that Mariner 9, whose cameras had a resolution in some areas of 100 meters—that is, they could make out objects about the size of a football field but no smaller—provided only a "modest" chance of detecting anything of lesser visibility. At the scientific conference just before the launch of Viking 1, Sagan showed a photograph, taken by Mariner 9, of what looked very much like a couple of pyramids at the limit of the camera's resolution; although he did not mean to suggest the existence of Martian pharaohs, he said, he did mean to suggest that as spacecraft went ever closer to the surface, achieving ever higher resolution, the scientists should be prepared for surprises.

For example, the Martian atmosphere, which Viking would have to go through, was a potential area for the unexpected, since NASA did not know exactly what was in it. Only recently, it had been ascertained by accident that the atmosphere might include the gas argon, a fact that caused some scientists to worry about whether Viking's landing sequence (the timing for the parachutes and the retrorockets) would be affected. Although it was not, the new information meant that the sequence of one of the experiments aboard the lander had to be changed. (The possibility of substantial amounts of argon was a good omen for life, for it suggested the atmosphere might have been denser in the past. As certain isotopes of argon that are outgassed early in a planet's life are too heavy to escape gravity and too inert to combine with other elements, they remain in the atmosphere and are an index of how much atmosphere a planet might originally have had.) The argon was suspected as the result of the erratic behavior of a vacuum pump during the descent of a Soviet spacecraft that ceased to transmit data upon landing; it was upon such fortuitous events as this that the character of the Viking mission sometimes depended. The most constant criticism of the project had been that it was trying to do too much too soon, that the exploration of Mars should have been conducted in slower stages. One Viking scientist said: "What you need is a rational strategy with steps. Why wasn't the first lander a hard one—that is, one that impacted on the surface and did something simple like measure the atmosphere on the way down? You need an orderly progression when you go to a new planet, as we did with the Moon: first you have a flyby mission; then perhaps an orbiter, followed by impacting a hard lander, which can sample the atmosphere; and *then* you have a soft lander. If we'd gone in steps, we'd know what was in the atmosphere of Mars now, so we'd know what we were doing—and not rely on prayer wheels. The agency could be in deep yogurt on this one."

Viking scientists were understandably nervous about unknown factors on Mars. According to NASA, the Russians had launched at least 13 craft toward Mars, though they acknowledged only 8. Of these 8, the ones known as Mars 1 and Zond 2 vanished without a trace. Mars 1 achieved orbit and a lander hit the surface, but was not heard from again (the Russians announced that it had taken a pennant to the planet—their way of putting the best face on a failure); Mars 3 landed successfully, but died after sending back featureless television data for only 20 seconds. Both these craft arrived during the huge 1971 dust storm, recorded by our Mariner 9, and because our craft could be reprogrammed (although it was only an orbiter with no lander accompanying it), it was able to send back by far the most data to come from any mission thus far. The Russian craft, being entirely preprogrammed, sent their landers uselessly into the storm while their orbiters transmitted pictures of a dust-concealed planet. Mars 4 and 5 were orbiters; the former failed to achieve orbit, and the latter was only slightly more successful; it orbited for 10 days and returned some 60 photographs of the surface. Mars 6 and 7 were simply "space buses" designed to carry landers to the planet. Mars 6 got within moments of landing, but when its retrorockets were supposed to ignite, contact with the spaceship was lost; this was the craft whose erratic vacuum pump led to the suggestion of the argon. The Mars 7 lander missed the planet completely. The failed craft represented a vast effort on the part of the Soviet Union. Mars 2 and 3, their orbiter-lander combinations, were thought to have weighed more than 10,000 pounds—substantially more than the Viking orbiter and lander, which together weighed not quite 8,000 pounds.

In the wake of the series of Soviet disasters, about all Viking had going for it was greater flexibility—something some scientists who had dealt with the Russians believed reflected a basic difference between the two countries rather than merely a difference between spacecraft. The landing sites would be monitored before lander separation by the Viking orbiters, and a lander would not be permitted to begin its trip to the surface until the site certification group—a team of scientists at the Jet Propulsion Laboratory in Pasadena, California, from which the craft were controlled—was satisfied that conditions on Mars were right. The group was not expecting any global dust storms, such as the one that greeted Mariner 9 on its arrival, because these seem to occur only when Mars, in its elliptical orbit, is closest to the Sun. Vikings 1 and 2 would be landing when Mars was farthest from the Sun. Smaller dust storms would be quite possible, though, and the group would be checking on whether there were any in the vicinity of the landing site. Whatever the planet might provide, Sagan had a couple of possible surprises in mind that were making life difficult for the site certification group. One was that winds of 600 or 700 miles an hour might blow up out of nowhere on Mars, at any place and at any time, to destroy a spacecraft during a landing. The chairman of the site certification group, Harold Masursky, the geologist who had suggested lowering Mariner 9's orbit, was frequently put off balance by Sagan's ideas about the planet, which tended toward the exotic; like most geologists, Masursky, a senior scientist at the Branch of Astrogeologic Studies of the United States Geological Survey, in Flagstaff, Arizona, was quite pragmatic about Mars. Masursky was once told by a colleague that he should have got a medal for running a committee—the Mariner 9 imaging team—that included both Sagan and Bruce Murray, with whom Sagan nearly always managed to disagree. Masursky was not without strong opinions himself. His specialty is volcanism, and years ago, before men went to the Moon, he used to argue that it had volcanoes, which proved not to be true; consequently, he regarded the discovery of the volcanoes on Mars, a concept that Murray had long held out against, as a sort of vindication. The three scientists were constantly at loggerheads. Masursky, a sardonic man with long, black hair and a thin, alert face, was highly amused when Murray, upon seeing Mariner 9's photographs of calderas—huge volcanic cones—on Mars, shifted nimbly from his original belief in Mars as a cold, dead planet to a Mars that was just waking up. "Having said all along that Mars was cold and dead, an awakening planet was the next best thing," Masursky said.

As for Sagan's notion that a 700-mile-an-hour wind might arise out of nowhere and zap the spacecraft as it landed, Masursky's eyes were apt to cross at the mere mention of it. The winds on Mars appeared to average between 10 and 20 miles an hour most of the time; though most scientists agreed that they could get up to 200 or 300. But because the atmosphere is so thin, a 100-mile-an-hour wind on Mars would feel like a 10-mile-an-hour wind on Earth, and Viking could land safely in a 150-mile-an-hour Martian breeze. But Sagan's 700-mile-an-hour wind would be more than enough to wreck a spacecraft during a landing. Vikings 1 and 2 would be landing at around 4 P.M. Mars time, however, a time of day when the winds were apt to be calm. Masursky once told Sagan that if the wind velocity was regularly as high as Sagan thought, Mars would long since have been worn down to the size of Phobos, one of its tiny moons. The wind speed had to be inferred from such things as the lengths of the tails of dust in the lees of the craters, estimates of the size of the dust particles, and the density of the atmosphere. "The data on the wind velocities are very squashy," Masursky said at the time. "If you take the most extreme circumstances—the smallest size particles, blowing about in the highest altitudes, where you have the rarest atmosphere—then you will get extremely high wind velocities." Masursky thought that Sagan sometimes arrived at his theories by picking up all the extreme cases and adding them up. The two scientists frequently disagreed at meetings of another committee they served on—the one that selected the landing sites. Masursky, who naturally wanted sites more interesting to geologists, almost managed to designate as a landing site a big highland volcano, Tyrrhenum Patera, which he said would be like landing on the edge of Mount Rainier. Sagan said that in the Mariner 9 photographs the dust streaks around the mountain were evidence of high winds there—whereupon Masursky pointed out that those streaks were the aftermath of the global dust storm that had been blowing 2.5 years before. Sagan, who is very good at thinking up epithets to stigmatize things he doesn't like, then began referring to Tyrrhenum Patera as "Masursky's kamikaze site," and that was the end of that. Sagan feels that in dealing with the unknown the outside chance should never be ignored.

A few years ago, he himself was guilty of ignoring a one-in-a-million shot, and he has never forgotten it. After a Russian spacecraft had landed on Venus in 1967, a Soviet scientist, Dr. A. D. Kuzmin, claimed that it had come down on a mountaintop, and Sagan replied that he did not think this was very likely. Kuzmin asked him what he thought the probability was that the first German bomb to land on Leningrad in the Second World War would kill the only elephant in the city's zoo. "I admitted that the chance was very small indeed," Sagan wrote penitently in *The Cosmic Connection*. "He responded, triumphantly, with the information that such was indeed the fate of the Leningrad elephant." Even though it was ultimately concluded that the Venus probe had not landed on a mountain-top, Sagan clearly didn't intend to be caught out again by any Leningrad elephants, on Mars or anywhere else.

At an altitude of 4,600 feet, 40 seconds before Viking touched down, its retrorockets would fire and its parachute, which had slowed the lander's speed of descent to about 140 miles an hour, would be jettisoned. In deference to any microbes that might happen to be at the landing site, Viking's retrorockets were not the single jet of flame upon which lunar modules used to ride down to the Moon; those seared the ground for yards around. Viking would ride down, at about 6 miles an hour, on three smaller rockets, each of which had 18 little jets, as though the rockets were shower heads. Consequently, any microbes in the vicinity, instead of being fried, would be subjected to little more than a warming wind. The fuel the rockets would use had also been refined in such a way as to minimize the effect of the descent. As the craft settled down, its three legs would compress to absorb the shock, pushing three round footpads slightly outward. Because of the lag between the time a signal was sent from Mars and the time it was received on Earth, the scientists and engineers at J.P.L. wouldn't know whether Viking had landed safely until more than 18 minutes after the event.

To Masursky's consternation, Sagan warned that the Viking lander might keep right on going into the ground and sink out of sight. A few scientists, Sagan vociferously among them, believed that the surface of Mars was strewn in places with soft and loosely packed dust that might not support the spacecraft—an idea originally promulgated in relation to the Moon, more than a decade ago, by Thomas Gold, the director of the Center for Radiophysics and Space Research at Cornell, of which Sagan is a staff member. The concept was one of a number of old ideas about the Moon that had been refashioned slightly and transferred to Mars. (In regard to the Moon, where there is a strong vacuum, the theory held that the deep dust had been deposited by electrostatic currents; on Mars the responsible agent would have been the wind.) Like Sagan, Gold had a flair for the dramatic or the improbable in science and was not apt to give up easily. Masursky, who was also involved with NASA during the trips to the Moon, recalled that at the time of Apollo 11, after several unmanned Surveyors had landed safely, Gold was still insisting that the astronauts ought to wear snowshoes and be equipped with avalanche cords on their helmets, so they could be pulled up if they sank in. While Masursky was going around to scientific conventions arguing that the Moon was hot and volcanic, Gold was following him arguing that it was cold and primitive. (Both men were proved wrong when the Apollo missions revealed the Moon to be neither actively volcanic nor completely inert.) Yet as recently as the summer of 1975, Gold was suggesting that the Martian calderas weren't volcanoes at all but pingos—mountains thrown up by underground deposits of ice. Though many scientists were frequently irritated by Gold and Sagan, they felt that the two had a special value. "They generate a vast quantity of ideas, a good many of which are wrong, but people are galvanized to disprove them," a geologist who knows them both has said. "Their ideas are so outrageous! Even more outrageous, of course, are the Immanuel Velikovskys and the Erich von Dänikens, who are so far out that few people consider them worth disproving." Velikovsky, a psychoanalyst who strayed into other fields, has argued that ancient catastrophes, such as the flood described in the Bible, may have been the result of planets straying from their orbits. Von Däniken has suggested that the Earth has been visited in prehistoric times by astronauts from other worlds. Sagan, in fact, is one of the chief critics of both.

The geologist went on: "Sagan and Gold, though, are within

the circle of those that you have to pay attention to and take the trouble to disprove, for what they say *could* be true, and in disproving them science is advanced. And there's always the chance that they might be right. Considering the number of ideas thrown out, if they were right just one time in a hundred that would be enough to give them a considerable reputation."

Most geologists didn't think much of the notion of deep, soft dust on Mars. "Those who think the spacecraft is going to sink out of sight are full of malarkey, but I can't prove it," one geologist said. "We geologists tend to be earthy about this matter," said Thomas Mutch, the rather diffident leader of the lander imaging team, whose mild manners are said to be sometimes subject to great strain. "It is difficult for me to conceive of these exotic fluffy materials that the astronomers and astrophysicists think up. I look at the pictures of Mars from Mariner 9, and I say the ground looks good. Sagan says that the resolution is poor—only one-hundred-meter objects show up. I say, 'Yes, I know that, but I can extrapolate from what I know of the Earth and the Moon.' He says, 'Yes, but can you really be sure?' And, of course, I can't."

Sagan, who couldn't either, nonetheless had a way of appropriating the unknown to his side of any argument. "Just because deep dust didn't turn up on the Moon doesn't mean it won't on Mars," he said, again using his favorite grammatical construction. Gold, who is blunter than Sagan and doesn't resort to double negatives, said at about the same time: "Deep dust on Mars? Of course Mars has deep dust." And Frank Drake, the director of the radio telescope in Arecibo, the instrument that has provided most of the radar readings for Mars, said forthrightly: "Some places on Mars turn out, under radar, to be loosely packed dust, and Viking *could* sink up to its eyeballs."

Even Masursky, who as chairman of the site certification group was one who would have to decide what weight to give the Cornell astronomers' warnings, admitted that when radar waves were bounced off Mars they give what he and everyone else called "odd reflectivities." Radar signals are like a lot of balls bounced off a surface: if the ground is hard and level the balls come bouncing straight back, but if it's soft fewer of them come back; the others get absorbed, causing what radar experts call "a diffuse echo." "The radar that was bounced off Mars at places near Chryse, the Viking 1 landing site, returned signals that were a little more diffuse than the readings we got on the Moon, so the softness of the ground *is* a proper subject to be worried about," Masursky said. There were other explanations for diffuse echoes; one was that the surface at the landing site was hilly or rough, so that a number of the hypothetical balls would bounce off at odd angles and not come back. This was the explanation Masursky and most of the other geologists favored. As it happened, the landing site for Viking 2, Cydonia, was too far north to be examined by radar from Earth, and Chryse, since the time of its selection in 1974, would not, until late June 1976—a week before the first Viking landing was scheduled—be in a position in relation to the Earth that would make a direct radar reading possible. Chryse had been bracketed by two radar readings, though; the one to the northeast gave a solid echo, but the one to the northwest gave a diffuse echo. Masursky's belief that the diffuse signals reflected not soft ground but rough terrain was based on the fact that in the Mariner 9 photographs of the area the terrain to the northwest showed many wind streaks and so probably consisted of sand dunes, which would give a diffuse echo not necessarily because they were soft but because they constituted hilly ground. Sagan, who was adept at taking the other fellow's model and making it look worse than his own, had said that hilly ground would be as dangerous as soft ground. Masursky replied that two of the unmanned Surveyors had come down safely on 19-degree slopes on the Moon. Viking had had successful landing tests on slopes of 25 degrees, and he didn't think it would have to contend with anything steeper on Mars. Later, when a direct reading would be possible, the site certification group "would work like hell," in Masursky's words, to analyze the new radar data before the go-ahead was given to Viking 1 to land at Chryse; data of this sort normally took 6 months to analyze. Masursky, who would be spending more and more of his budget on the questions raised by the Cornell astronomers, would use the radio telescope at Arecibo and another at Goldstone, California, to make the necessary rapid analysis. Sagan wasn't a member of the site certification group. Some scientists felt he was so inventive that he was not the best person to have around when options were being narrowed. "Carl has a way of expanding them," one scientist said.

If the Viking 1 lander didn't sink up to its eyeballs, several things would happen as soon as it landed. A big, dishlike radio antenna would pop up and swivel toward the Viking 1 orbiter, which, with its more powerful transmitter, would handle the lander's communications with Earth. A short arm, with a boxlike weather station at the end, would extend to one side of the lander. And exactly 24 seconds after touchdown, by which time the dust (if there was any) kicked up by the rockets should have settled, one of two cameras perched side by side on top of the lander would start taking the first picture of the surface. The Viking lander was a low, sleek, squat craft, designed to be proof against the winds and storms of Mars. It was painted a battleship gray, and like a battleship's, its superstructure was covered with odd bumps and turrets, which had been smoothed in order to present no rough edges to the wind; they screened fuel tanks and instruments. Viking's main hull was a 6-sided aluminum box,

18 inches deep. Its sides were alternately 43 and 22 inches, and it was held 9 inches off the ground by the three legs. The two gray cameras looked like a pair of funnels on a ship; in front of them, where they would be able to photograph what it was doing, was the main sampler arm—a shovel on an extendable pole, which at touchdown would be coiled inside its housing. Between the cameras, like ventilators on a ship, were three gray hoppers, into which the sampler arm would later drop Martian soil. The hoppers would convey the soil below decks, where three main instruments would analyze it: the X-ray fluorescence spectrometer, which would assay the soil for inorganic material; the gas chromatograph mass spectrometer (GCMS), which would analyze the Martian atmosphere and also play a part in the search for life by looking for organic compounds (those based on carbon); and the biology package, which was itself divided into three instruments for detecting living microbes. Behind the cameras was a seismometer, housed in a low rectangular box.

The scientists at J.P.L. would be eager to get the first picture back to Earth quickly, before the orbiter, which relayed it to Earth, got out of the lander's range; since the only Soviet craft that made a successful soft landing on Mars ceased to send signals 20 seconds after touchdown, it seemed that speed might be important, although the cameras worked very slowly. They were of a type known as facsimile cameras, which the Russians had used in their lunar and Venus landings. The lenses were behind long, narrow slits; a mirror inside each slit moved up and down, reflecting the view through the lens line by line onto an array of light-sensitive diodes as the camera slowly revolved; later, at J.P.L., the picture would emerge in vertical strips, a little as if a man were papering a billboard. (If dust got on the glass pane covering the slit, a small jet of gas could clean it off, and if dust got beneath the camera, so that it was unable to swivel, a tap from the sampler arm could set it going again.)

The first picture would be just a small patch of ground around one of Viking's front footpads. Immediately after that, a second picture, a panorama of the horizon, would be taken; both would be in black and white, which could be transmitted faster than color; on the following day, or *sol*, as the flight controllers call a Martian day, the cameras would start photographing in color, and 2 days later they would begin taking stereoscopic pictures. Because the sunlight on Mars is much weaker than that on Earth, the quality of the color photography might well be poor for the first few days, until the engineers at J.P.L. adjusted the cameras to the unusual light—something they would do with the aid of a color chart on a windscreen behind one of the cameras. On the windscreen behind the other camera was the undulating star of the Bicentennial emblem. As the cameras—which were mounted low on the deck to take advantage of the spacecraft's warmth—rotated from color chart to Bicentennial emblem, their view of Mars would be cut into by all the different knobs, funnels, and hoppers; consequently, the first view of Mars would be a furtive one, as though a scout were peering out at an enemy from a gun turret.

Though most of the scientists said that as soon as they got a chance to look at the first photographs from Mars they would search the pictures for any sign of life, many were prepared to give up the quest not long after they had scanned what they expected would be a flat, barren-looking landscape. Sagan, though, would continue to examine the photographs for indications of living organisms. "*Someone* should look for these things," Sagan said. Since the Viking cameras could not detect microscopic life forms, Sagan would be looking for what he called *macrobes*. This was a term he once coined to describe any life visible to the naked eye (as opposed to microbes, which are not), and since the cameras have about the same resolution as the eye, he found the term useful to describe what he would be looking for on Mars. One of his major complaints about Viking was that it was not mounted on wheels or tractor treads (something that apparently could have been done at little extra cost if it had been thought of early enough). "I keep having this fantasy that there, twenty feet away, will be this tree or something, and we can't get to it," he said at the time. Lederberg shared this concern, for he believed that if the Martian deserts were anything like Earthly ones, and had vegetation on them, the vegetation would be in the form of widely scattered bushes and oases. Still, the trees and bushes—if they existed at all— might well be invisible to even completely mobile cameras because a model for Martian trees that was popular with some biologists (though not with Lederberg) placed them underground, to avoid the ultraviolet rays of the sun and the extremes of temperature, with long taproots descending to a remote water table or permafrost, with perhaps just a flat segment at ground level, like a manhole cover, which could absorb energy from the Sun.

One of Sagan's arguments for the possibility of macrobes on Mars was that in a cold, dry climate, where an animal must retain warmth and moisture, the bigger he is the better, because a larger creature has less surface area in relation to volume, and, of course, it is from an animal's surface that heat and moisture are lost. "Organisms with an interest in the conservation of heat and water thus may select larger sizes. [This is one reason why there are large mammals preferentially in the terrestrial Arctic]," Sagan wrote in another of his books, *Other Worlds*. Next to these words is an aerial photograph of a polar bear sauntering across a snowfield, where he is almost as difficult to spot as he would be on a Martian ice cap—suggesting, the inference may be, why similarly protectively colored macrobes on Mars (if they existed) had not been detected so far. *Other Worlds* is a mixture of text, photographs, drawings, and poetry whose overall purpose is to incline the reader toward a greater receptivity to extraterrestrial life—an idea whose time, Sagan felt, had come. Where the black type covers the white snowfield and even impinges slightly on the back of the polar bear, he wrote, first allying himself with opponents of excesses such as those of Percival Lowell:

> Some people are nervous about the possibility of life on Mars, even simple life. There have been so many premature and unsupported speculations on this topic that some scientists and newspaper editorialists often take the most conservative viewpoint they can imagine. They are sometimes found supporting the curious principle: "When faced with two alternatives, in a subject about which we know nothing, choose the least interesting."

No reader of *Other Worlds* is ever likely to make such a mistake. After pointing out that too little was known about Mars to exclude anything—a proposition normally used to make the opposite point—Sagan continued:

> When asked to picture what will be uncovered by the first few Mars landers, many scientists will admit the possibility of microscopic organisms [microbes], but will deny vehemently the possibility of macroscopic organisms [macrobes]—that is, beasts discernible to an unaided human eye on the surface of the planet.

Though many scientists did indeed admit the possibility of microbes on Mars, they also argued that Mars is so cold and so dry, and its atmosphere so rarefied, that the possibility was not great; consequently, their lack of enthusiasm over the possibility of macrobes was actually less of a matter to be wondered at than Sagan has implied. Sagan passed quietly close to, if not exactly into, the realm of Lowell:

> In fact, there is no reason to exclude from Mars organisms ranging in size from ants to polar bears. And there are even reasons why large organisms might do somewhat better than small organisms on Mars.

Not quite into the realm of Lowell, of course, because Sagan, by means of the double negative, had never once committed himself to the proposition that large animals do exist on Mars; he had merely held it up as an enticing (and easily retractable) possibility. Most scientists didn't think that Sagan bore much resemblance to Lowell. "He is far too sophisticated a scientist, for one thing," a colleague has said. "But if Lowell hadn't existed, Sagan might have invented him."

The reason most scientists were willing to exclude the possibility of macrobes on Mars was that on Earth organisms large enough to be seen are a result of perhaps 4 billion years of luxuriant evolution—an abundance not even Sagan claims for Mars—and, even so, macrobes didn't appear on Earth until 600 million years ago, or within the last 15 percent of this planet's extremely active life span. For almost 3 billion years before that, life on Earth had consisted of single cells that lived entirely in water; life apparently required a long incubation period in the ocean—a condition highly unlikely on Mars—before it evolved into multicellular forms. Lederberg, who had a hand in the original conception of the macrobes, did not appear to put as much weight on the idea as Sagan. "I don't know which of us arrived at the idea first, but it started with the notion that the harsh conditions on the surface of Mars might make it hard for small things, so big things were more likely to flourish there," he has said. "It's hard to imagine life starting off with a bang with, say a giraffe, but if conditions on Mars were more hospitable long ago small organisms might have evolved into big ones. And in the Arctic you do have polar bears." However, Lederberg sounded as if he were simply trying to stretch his imagination in order not to be accused of being too Earthbound in his thinking—an exercise in mental gymnastics rather than a statement of belief. Sagan and Lederberg by no means always agree; Sagan has said that the older scientist is harder-nosed and less romantic than he is himself. Other biologists are less charitable about the macrobes. Norman Horowitz, of the California Institute of Technology, who was a member of the Viking biology team, has pointed out that although there are polar bears in the Arctic, there are big animals in Africa, too; and conversely, in the coldest, driest regions of the Earth— certain valleys of the Antarctic—there are no higher forms of life at all. The last survivors in such areas are microorganisms, he has noted, which, with respect to cold and dryness, are the most adaptable, and in the very coldest Antarctic valleys, which are a paradise compared with Mars, even these are probably dormant. Some of Sagan's colleagues have said he lives in a dream world. "He's charming, he's bright, but he's totally uncritical," one scientist said at the time.

If Sagan was pressed, he would admit that the concept of macrobes on Mars wasn't a very likely one. Some of his colleagues said he was overstressing it mainly with an eye to adding a bit of excitement to Mars, Viking, and the entire space program. Certainly, NASA valued his public relations abilities highly, as well it should have, and sometimes consulted him on matters concerning the press. Sagan found NASA's publicity on Viking rather lackluster—the agency played up the nuts and bolts of the spacecraft instead of emphasizing the more dramatic qualities of space exploration, he said—and this deficiency was one that he, intentionally or not, rectified almost single-handedly. At the time of the Viking launch, a friend of Sagan's said he seemed to be undergoing a sort of conflict between being a scientist and being a popularizer of science, whether by writing about it or talking about it. (Sagan wore two NASA badges to the launch, one identifying him as a scientist and the other as a correspondent for *Icarus*, the scientific journal he edits.) Sagan sometimes wondered whether he spent too much time on the lecture-and-television circuit; he was alarmed when his son Nicholas told him he wanted to be two things when he grew up—"a daddy and a host," meaning by the latter the host of a TV talk show such as he has frequently seen his father appear on. His exposure on radio and television was the source of more criticism from other scientists than anything else he did. "Hell, Carl doesn't have any compulsion to find life; life's the most pizzazzing idea around, that's all," one scientist protested at the time. Another of Sagan's Viking associates said: "It's hard for many of his colleagues to see his appearances on the Johnny Carson show or the 'Today Show' as justified by science; they feel that popularizing science is incompatible with *doing* science. I don't feel that to be true myself." He added, after a moment's reflection: "A certain amount of jealousy enters into their judgment. Heck, we'd all like to be on the Johnny Carson show."

Some of Sagan's associates profess themselves to be glad he is doing what he is doing. "He is a very sophisticated public relations man, and he gets the issues before the people," one said recently, "and I'm delighted that he is spending his time this way." Sagan himself has said that the kind of science he likes to do requires money—the public's money—and someone has to take the scientists' case to the people. He is very effective. After a television appearance in which he talked of the possibility of life on Titan, the largest moon of Saturn and the only satellite in our solar system that is known to have a substantial atmosphere, Congress approved money to include it on NASA's next planetary mission. (It was launched a year after

Viking, reached Jupiter in 1979, and would reach Saturn and Titan in 1981.) Some of his colleagues felt that Sagan was especially valuable to NASA now that the excitement generated by the Apollo program had abated because he kept the possibility of extraterrestrial life in the public eye. Viking cost $1 billion, of which over $100 million went for the instruments to detect organic compounds and life. Many people felt that the science had got fouled up as a result; if there were no Viking mission, they argued, there wouldn't be such a tremendous effort to make Mars appear habitable. They likened the situation to the construction of an expensive cyclotron designed to look for a certain particle: the builders have to maintain a belief that that particle is there because they'd be in an awful fix otherwise.

Sagan would most likely have thought of the macrobes on Mars even if there had been no extraneous reasons. "I think he's sincere," said one scientist who has known him for years. "He really calls it the way he sees it." The macrobes were a model Sagan may have taken less seriously than his audience, however; some of Sagan's models may simply have been metaphors for what might be. He believed that conditions on Mars, let alone on some of the more remote planets, are so different from those on Earth that if life existed there it would be bound to take forms we consider unlikely. Therefore, he maintained, instead of restricting our imaginations, we should let them go.

Before Viking, nobody, of course, could say there *weren't* any macroorganisms on Mars, and the possibility of their existence—surely not to be discounted in other solar systems regardless of the outcome on other planets of this one—presented curious problems of perception that could conceivably arise again. In Sagan's opinion, the first problem with macrobes would be to recognize them, for surely they would not look like anything known on Earth. This, as Sagan pointed out, was certain because any example of life on Earth is the result of so many thousands of random, almost accidental turns during its evolution that the chances that any form of life on another planet would follow an identical path are virtually nil; it couldn't even happen on Earth if this planet were to start all over again. "Never again, in any other place, will there be people or cows or grass," Sagan said at the time. The individuality of life on Earth is the only respect in which many scientists today regard that life is unique. Consequently, if a rabbit, say, turned up on the photographs from Mars, the imaging team would be certain that it got there somehow from Earth, and, conversely, an Earthling might stare right at a photograph of a Martian macrobe and so different might it be from one of our own that he would not know it was something alive.

Sagan thrives in totally novel situations like this; they offer fascinating possibilities. They necessitate a definition of what one is looking for—which in this case was life itself. Oddly, there had never been a time when the Viking scientists sat down and asked themselves what life is and what the best ways to look for it are—at least, not since the conference that the Space Science Board of the National Academy of Sciences sponsored in 1964 and 1965, which examined the prospects for biology on Mars. The biology package itself came into being in an ad hoc way. Most biologists admit that the definition of life has become increasingly elusive since the days when a living object was tentatively described as one that did such things as eat, drink, metabolize, breathe, grow, reproduce, and possibly move. "All those old definitions don't really fit," Sagan said. "My automobile eats and breathes and metabolizes and moves. Crystals grow and even reproduce." Though Sagan may be quite freewheeling about life, especially extraterrestrial life, he has probably thought more about its attributes—with the aim of devising a definition useful for identifying it, wherever it may be found—than anyone else ever has. He wrote the article under the heading "Life" in the most recent edition of the *Encyclopaedia Britannica*, and it is probably the most lucid account of current thinking on the subject; in it he states that evolution by natural selection is the feature that distinguishes living matter from nonliving. He has said: "There is no other foolproof way to get at the complexity of life, for most biologists have come to the conclusion that the only certain thing they can say about life is that it is something that *evolves.*"

Unfortunately, Viking scientists couldn't sit around for thousands of years waiting for macrobes to evolve in front of the cameras, so Sagan had to find another criterion for examining the photographs Viking transmitted from Mars; a criterion he settled on is a relatively new one—*thermodynamic disequilibrium.* This is a theory that argues that any major departure from the most likely chemical or physical situation may be attributable to biology. "Living things frequently depart from thermodynamic equilibrium," Sagan said. "A tree, as long as it is alive and growing, will stick up in the air because its roots will hold it up; sooner or later, though, after it dies it will be flat on the ground. The same is true of a cow, which is heavy on top and is held up by four spindly stilts; it should be flat on the ground, and sooner or later it will be, but as long as it isn't you suspect it's alive. If it were nonliving, a rock shaped like a standing cow, for instance, gravity plus erosion would collapse it. If we see a herd of cow shapes on Mars, I'll bet they won't be rocks, but some kind of animal. At least if they were rocks they'd have to be of relatively recent origin—perhaps artifacts *produced* by some kind of animal. There is a wide variety of disequilibriums, and they're easy to recognize. Suppose a fourteen-foot squamous purple ovoid with thirty tentacles came floating through the air, and Viking got a picture of it. We'd identify it as alive even though we'd never seen it before and didn't know its chemistry, simply because it was so improbable."

Improbability, though it may be a quality of life, is no surefire test for it. Thomas Mutch, the leader of the imaging team, who would be using the camera to study not macrobes but the terrain and any geological features, and who, like most geologists, didn't think much of the chances for any sort of life on Mars, liked to display a photograph of a group of rocks called *ventifacts*, which he believed could well turn up in the pictures from Mars because they are peculiar to windblown deserts. What is more, if they did, he and Sagan might well find themselves in strong disagreement about them because ventifacts are unlikely looking objects in distinct thermodynamic disequilibrium; they have been carved and abraded by windblown sand, and because the heavier and more abrasive bits of sand blow along closer to the ground, the rocks have been worn away to a greater extent at the bottom, so they are distinctly top-heavy. Moreover, they have been worn by the wind into weird and wonderful shapes—surreal triangular faces emerging from the ground like Easter Island sculptures, perhaps—and because they are frequently found in large groups, like the Easter Island statues, they could well be taken for a delegation of curious Martians come to greet Viking.

If Sagan and Mutch had any doubts about whether an object was an animal or a mineral, there was another test they could apply to the potential macrobe, and that was motion. The lander's cameras were still cameras, not movie cameras, but an object that was present in one picture and absent in a later one, or that had changed position, would be known to have moved, though whether it had moved itself or been moved by the wind would not be immediately clear. With luck, the cameras might catch something in the act of moving. When the imaging team tested one of the cameras in a particularly Martian-looking stretch of Colorado desert in the Great Sand Dunes National

Monument, a few hours' drive from the Martin Marietta Aerospace plant in Denver where the lander was built, Sagan rented from a pet store a garter snake, a chameleon, and two tortoises, which he let loose in front of the slowly rotating optics. If the reptiles were moving in the same direction the camera was turning, they became elongated; whereas if they were moving in the opposite direction, their images were compressed. (The chameleon was too fast for the sedately moving mirror inside the camera to capture, but as it skittered away it left in the sand a handsome set of tracks that Sagan claimed would speak volumes if they were transmitted from Mars.) When the scientists posed for a joint portrait, they were able to leapfrog to the right just ahead of the camera's movement so that their images were repeated over and over and they looked like a crowd of students posing for a class photograph. Several Carl Sagans crouched in front, and several Thomas Mutches, all of them tall and spindly and wearing glasses, loomed in the background.

Sagan was dissatisfied with the cameras' ability to detect motion on Mars, since the most important evidence depended so strongly on a macrobe's being at the right place at the right time. He was pleased by the addition of a feature called a *single-line scanner*, which enabled the cameras to repeat the same swath over and over again. That way, anything that moved past would be caught as a blur against a constant background, the cameras recording the motion, though not the object itself, the way a man peering through a slit in a fence might see the blur of a car going by. The value of the single-line scanner depended on how often it nodded up and down in the same place, just as the likelihood that a man looking through a crack in a fence will see a car whiz by improves the longer he watches. Though the lander was supposed to be operated from J.P.L. while it was on Mars, a program of events covering its activities during its first 2 months on the surface would be fed into its computer ahead of time, in case J.P.L. lost control of the craft after touchdown. Sagan had lately been annoyed because the number of preprogrammed single-line scans had continually been cut down; Viking's abilities were strictly limited, and the scientists had to compete for its time. "Carl wanted to rev up the camera in order to see rabbits hop by," a geologist said with more than a trace of consternation. "I had to admire Mutch for not shooting him. Carl's the greatest menace since the black plague." Up until the launch of Viking 1, the geologists didn't see a great deal of Sagan, but whenever he had anything on his mind he managed to keep in touch by mail. His letters tended to be a bit peremptory. When he heard that the single-line scanner would repeat its action only 30 times when it was operating in its preprogrammed mode, he dashed off a five-line note to Mutch saying that 30 repeated scans were too few, that the situation alarmed him, and that he would appreciate a clarification. "Carl is a great provocateur," Mutch said one day in his office at Brown, tossing the letter into a drawer. "He's brilliant. Sometimes he's right, as he and his associate James Pollack were about the seasonal patterns on Mars being caused by dust blown at varying wind speeds, and so on. And he may turn out to be right about life in space, which would be his great vindication."

Before the launch of Viking 1, Sagan made one of his rare appearances at a meeting of the lander imaging team at the Kennedy Space Center—primarily because he had heard that the preprogrammed sequences, including the use of the single-line scanner, would be discussed. He arrived 45 minutes late and took an empty seat next to Mutch, who was chairing the meeting. Sagan is almost always late—a habit that infuriates his colleagues. He never wears a watch, and he has explained: "When I'm doing something I like, I hate to break it off because

I have to go somewhere else. It's more fun to keep on." As soon as he took his seat, a telephone on the table in front of Mutch rang; the call was for Sagan, and he spent the next 5 minutes deep in conversation.

Sagan had arrived in the middle of a presentation by Terry Gamber, an engineer from Martin Marietta Aerospace. (Engineers have more control over the cameras than the scientists have, and during the meeting the scientists professed to find it difficult to communicate with them.) Sagan asked about the status of the single-line scans, and Gamber told him there would be 20 or 30 repeated swaths taken after some of the photographs had been completed. Gamber added that the repeated swaths had been included as extra margin—that is, in the event the whole picture couldn't be put on the spacecraft's tape recorder for transmission to Earth, they would be dropped. Sagan was irritated. "But on days you don't have that problem?" he asked. (Though his manners are often courtly, they can verge on the imperious.) Gamber replied that even then there was no guarantee that the single-line scans would be transmitted from Mars; other data might have higher priority. (As the instruments gathered more information than could be transmitted to Earth, the scientists would be competing to get their data back.) Sagan tensed and asked whether the engineers who were running things had any idea of the importance of single-line scans. Gamber said he had told them. "Any chance we can make the arguments personally?" Sagan demanded acidly.

"We'd all like to do that," said Mutch, moving in to smooth things over. "But I don't think the engineers are wallowing in ignorance about what we want to do, Carl." For the remainder of the meeting, Sagan put in probing questions every here and there, often about the single-line scans. Were any of them to be in the afternoon or the evening, for different lighting conditions and possibly different types of movement? Were there to be single-line scans of some dust placed on top of the lander, which the wind might move? When James Pollack, his former student and now an associate of the imaging team, proposed that the cameras be used to photograph the Martian moons, which move visibly across the sky, and also the planets, including the Earth, Sagan was quick to point out that the latter idea would have enormous public relations value for NASA. (From Mars, the Earth would be a bright blue dot that, because it is nearer the Sun, like Venus, would have phases and be the brightest "star" in the morning or evening sky.) Sagan's mind, so resourceful in the early phases of an undertaking when options were still open, was constantly on the alert for anything his imaging team colleagues might have missed. At one point he seized on a matter that was so simple no one had spotted it before: concerning the small jet of gas that periodically dusted the camera lens, as though a person were blowing a speck out of his eyes, Sagan asked whether anyone had bothered to check with the other scientists whose instruments were analyzing the Martian atmosphere to see if they were aware of this possible contamination. No one had checked with them. "Otherwise, it might be a month later that someone noticed a correlation between his marvelous discoveries and our dusting," Sagan said. Mutch advised the engineers present to look into the matter.

In spite of the inclusion of the single-line scanning technique, Sagan continued to worry about the cameras' ability to detect life. "I keep having this recurring fantasy that we'll wake up some morning and see on the photographs footprints all around Viking that were made during the night, but we'll never get to see the creature that made them because it is nocturnal," Sagan said. "I wanted a flashlight on Viking, but I didn't get it." Mutch had tried to get Sagan his flashlight. Though it might not illuminate nocturnal macrobes, Mutch thought, it could still show the formation of frost, which he was interested in. He was unsuccessful, in part because any change at all in Viking, no matter how small, always seemed to wind up costing $100,000 just for the paperwork, and also because, with only two 35-watt generators to operate the lander, there wasn't enough electricity. "I don't really feel too cut up about it," said Mutch, who has a relaxed, gently mocking manner, which colleagues consider useful in curbing Sagan's irrepressibility. He went on: "Carl, by the way, also talked about putting out bait." This was a proposal to paint the outside of the spacecraft with a variety of nutrients, to see which ones the Martians liked best, so that if NASA sent another craft to Mars it could dispense the right foods.

Even Lederberg doubted whether a flashlight would be much use in detecting life. "Since it's so cold on Mars, I don't expect anything to be nocturnal; it would be coldest at night, and most macrobes would be asleep then, trying to conserve their warmth," he said. "I don't know how serious Sagan is about this anyway."

Sagan said he *was* serious about the flashlight, and he took issue with Lederberg on the matter of Martian macrobes all being asleep at night. "Some animals on Earth, such as Arctic fish, have elaborate antifreezes, and some organisms even have glycol, so that they can survive in temperatures five degrees or ten degrees below freezing, and there are even lower antifreezes—insects, I believe, have the lowest," he said shortly before the launch of Viking 1, piling one extreme life form on top of another in order to come up with a model of a nocturnal Martian macrobe. "Now, if you're a Martian organism and you don't want to freeze at night, even without antifreeze you could still heat yourself metabolically. And if you're a predator, what better device would there be than to go around at night with a heat sensor, eating Lederberg's sleeping organisms? If there are organisms sleeping at night, they're sitting ducks for predators.

Yes, I can imagine such an ecological function." (Sagan seemed annoyed that Lederberg was too hard-nosed to go along with nocturnal macrobes, which may be one reason Sagan's macrobes devoured Lederberg's somnolent ones so voraciously.) Most biologists, however, pointed out that though there may be some organisms on Earth that have adapted, with antifreezes and other methods, to extreme cold, they can hardly be said to have taken over even our Arctic tundras, let alone our Antarctic valleys.

Sagan was asked whether he considered himself to be filling his own special ecological niche as a terrestrial macrobe for stretching other terrestrial macrobes' consciousnesses. "Someone has to propose ideas at the boundaries of the plausible, in order to so annoy the experimentalists or observationalists that they'll be motivated to disprove the idea," he said. "Otherwise, there is a powerful temptation for an experimenter to design experiments for just what he knows. Of course I've annoyed the observationalists. But there would have been no search for large forms of life on Mars, or study of the variable features there, if I hadn't pushed for them."

Nothing very dramatic would happen aboard Viking until the eighth day—or eighth *sol*—after it landed. (The Viking scientists had to think in terms, not of days, but of Martian sols, which are 39 minutes longer.) Then the surface sampler, a smaller and more complicated version of the bucket of a steam shovel, complete with jaws and teeth, would emerge from its housing in front of the left-hand camera. The sampler would extend outward and slightly upward on the end of a boom that lengthened in a seemingly miraculous fashion, for the box looked far too small to have contained the boom's ever-increasing length. The boom would appear to *grow*, almost as though the box were a mill for making metal rods. Actually, the boom worked like a steel tape measure; there were two ribbons of prestressed steel, which unrolled in such a fashion as to form a single tube with a somewhat oval cross section. The boom had to have a long reach: the craft's landing rockets, despite their refined fuel and softened thrust, might have contaminated the area immediately around the lander, and therefore samples had to be taken from as far away as possible. At full extension, the boom was about 10 feet long.

The first few samplings had been preprogrammed, but on later ones, if a patch of different-colored soil or any other oddity that suggested life turned up in the photographs of the ground in front of the lander, the sampler would be commanded to head for that. When the boom was properly extended, the housing would swivel so that the sampler was over the target; then the boom would lower it to the ground. A sensor in the shovel would shut off the motor as soon as it touched, for with the 37-minute delay in the time for round-trip radio signals, the boom, of course, couldn't be operated from the Earth. The shovel's jaws opened, and the boom pushed out, straight away from the spacecraft, as though a man were digging with a spade. If the soil was sandy, the shovel would penetrate several inches into the ground. Then the jaws would shut; the boom would retract, elevate, and retract again as it began to recede into the box, which would swivel when the shovel was almost completely retracted. When the shovel was over one of the three hoppers, the jaws would open, and the material would disappear inside the lander.

On the first day of sample gathering, the shovel would perform its task three times—once for each hopper. The first hopper to get soil would be the one for the three biology instruments, then the one for the organic chemistry experiment, and finally the one for the inorganic chemistry experiment. (Space scientists, with typical brevity, use the word *experiment* to designate both the apparatus and its function.) The inorganic chemistry experiment was added very late in the development of Viking, largely at the insistence of Masursky and other geologists, who felt that the chances of detecting life on Mars were so remote that it would be wise to give the spacecraft additional tasks to perform. Determining the elements present in the soil of Mars was one of a number of basic experiments some scientists felt should have been done before any attempt to look for life. When the Martian soil trickled into the inorganic chemistry hopper, it would be channeled to a compartment where it would be bombarded with X-rays, which would stimulate the elements in the sample to emit their own rays—to fluoresce—at their characteristic wavelengths. In this fashion, the instrument could identify some 80 elements (although, of course, nowhere near that many were expected to be detected in any one sample). The geologists in charge had only 2 years to bring the instrument from prototype to final design, and they felt they could have done better if they had had more time and more money. The fact was that the physical sciences, such as geology and meteorology, had been downgraded in relation to the biological ones; the seismometer and the meteorology experiments had low priorities when it came to money and to the allotment of radio time for transmitting data back to Earth. The geologists' inorganic chemistry experiment, although it was better off than those (partly because it did not require much radio time to transmit its data), still had a lower priority than the organic chemistry or biology experiments. The geologists felt a little like second-class passengers in relation to the biologists.

The soil that disappeared down the hopper for the organic chemistry instrument—the gas chromatograph mass spectrometer, or GCMS—would be analyzed for organic compounds.

First, the soil would drop through another sieve, and then it would go through a grinder, where it would be pulverized before it moved on into the instrument itself. There the dust would be heated so that any organic compounds would be expelled; these would pass in a stream through the gas chromatograph—a coiled tube that allows the smaller molecules to pass through faster than the bigger ones, sorting them by size. After that, the molecules, one size at a time, would pass on into the mass spectrometer, which would tell what they were. Most Viking scientists believed that the probability of finding organic compounds on Mars was quite good; Klaus Biemann, an organic chemist from the Massachusetts Institute of Technology and the leader of the team in charge of the instrument, the molecular analysis team, rated the chance as high as 50 percent. The result of the collective work of men like Alexander Oparin, H. J. Muller, Stanley Miller, and Lederberg has been to provide an understanding of the continuous chain evolving from elements such as carbon and hydrogen, through organic compounds and single-celled organisms, up to intelligent life. If Darwin's theories needed further certification, these men have given it; they have shown how evolution works on Earth by taking it all the way back to the most rudimentary compounds, and they have paid it the added compliment of projecting it to include life everywhere. Since most Viking scientists viewed extraterrestrial life, and in particular, carbon-based extraterrestrial life, as inevitable, they were preoccupied with the narrower question of whether Mars was the kind of place where that sort of life could occur. They knew, for example, that the Moon could never have developed carbon-based life because it lacks, and apparently has always lacked, the atmosphere necessary for the formation of the two essential components of the living cell—amino acids (which combine to form protein) and nucleotides (which combine to form nucleic acid, the genetic material). However, molecular biologists thought it possible that life might have got a start on Mars, whose atmosphere may once have been more substantial than it is today. There could be quite complex organic compounds on Mars, the biologists felt, and if no higher forms of life developed to gobble them up they should still be there. (Though ultraviolet light is, of course, harmful to organic chemicals, a millimeter of dust would be sufficient to protect them; while the wind might blow the dust away in some areas, it would pile up the dust in others, and meteoric impacts were continually bringing material from deeper down to the surface.) The next step in the evolution of organic life would be for the nucleotides to polymerize—that is, to link themselves together to form chains of nucleic acid, which are capable of duplicating themselves. The chains contain genetic information that can be passed on from one generation to the next. The polymers of nucleic acid also direct the formation of tremendously complex proteins from building blocks of amino acids. It is this complexity that convinces most exobiologists, including Sagan, that life in the universe must generally be based on carbon, for no other atom has the same versatility when it comes to bonding with other atoms—not even silicon, which is the closest runner-up and which used to be thought of as a possible alternative. Sagan continually warns against the chauvinism (one of his favorite epithets) of making extraterrestrial life in our own image, and though he acknowledges that the consensus about carbon just could be pure chauvinism, he nonetheless dismisses silicon, saying: "You cannot make stable silicon molecules complex enough to contain anything like the information contained in the polymer chains of our genetic code. We are, of course, biased toward what we're made of, but when I try to think of other elements as a basis for life, I always wind up being what I call a carbon chauvinist." Though Sagan didn't rule out the possibility that somewhere there might be life based on silicon, or on something else entirely, he saw little reason to suppose that this was generally true, especially since carbon is abundant all over the universe, as are the other elements on which our biology is based: hydrogen, nitrogen, oxygen, and phosphorus. And there was no reason at all to suppose that the biology of Mars was chemically different from the biology on Earth.

On Earth, the development of the long polymers occurred in the oceans, which, as the strands duplicated and reduplicated, became a rich organic soup. Mars, though, seemed never to have had large oceans—at least, not for long periods of time. This was something the geologists delighted in calling to the biologists' attention; however, Mars might well have had smaller bodies of water, comparable perhaps to Earth's inland seas, for shorter periods, and this might have been enough to initiate the formation of nucleic acid and proteins. Indeed, Lederberg and others felt that shallow ponds could have been more apt than oceans to absorb the energy from the Sun necessary to generate the chemical reactions that produce polymers. (Generally speaking, the geologists, who were pessimistic about life on Mars, know most about the physical conditions on the planets; but the biologists may have the last word concerning planetary life because they know more about the conditions under which life could develop, and these may well turn out to be quite broad.) The GCMS would not be able to detect long polymer chains directly because the instrument heated the sample so that any chains would be broken down; yet Biemann thought he might be able to identify the chains—if they were there—from their components. Of the instruments looking for signs of life on Mars, many Viking scientists felt Biemann's might prove the most useful. After Viking landed, though, there would be less interest in that instrument than in what happened to the soil that dropped into the hopper leading to the biology package—undoubtedly the most intricate collection of remotely controlled instruments ever made.

The biology package was operated by the biology team, a group led by six principal scientists, including Lederberg, who would be testing for the presence of microbes—not macrobes—on Mars. Sagan was not a member of the team, and some of those who were became nervous whenever he appeared on television or gave interviews discussing life on the planet. They sometimes accused him of taking what we don't know about life in the universe and working downward to what might be expected on Mars, whereas they thought of themselves as taking what we do know about life on Earth and working cautiously upward—a way of thinking that Sagan ridiculed as chauvinist. Sagan's relationship with the professional biologists was complex and equivocal, for although he himself is not a biologist but an astronomer, he knows an enormous amount of biology and is intimately acquainted with a number of outstanding biologists. Moreover, Sagan and most biologists agreed on a great many points about extraterrestrial life, such as that it is based on carbon and that it needs water. They agreed that life on Mars would be jeopardized by the Sun's ultraviolet rays, which break the bonds that keep organic compounds together; these rays reach the surface of Mars because the Martian atmosphere, less substantial than ours, apparently contains little or no ozone, which filters them out on Earth. (Many biologists had proposed ways that Martian life might have developed to circumvent this hazard, such as the possibility that it existed underground or grew some sort of shell or even exuded its own suntan lotion. Sagan, of course, was especially enthusiastic about such possibilities.) The most basic agreement between Sagan and these colleagues about extraterrestrial life was that it existed.

Indeed, with so many stars in the sky and so many organic compounds among them, the odds are that somewhere, somehow, many of Sagan's, and other people's, more spectacular ideas will prove correct. So it might not be important whether squamous purple ovoids or other macrobes turned up on Mars. They might well exist somewhere else—or rather, as exobiologists say, there was nothing to exclude their existence elsewhere.

Although Sagan is a self-confessed "carbon chauvinist," and even a "water chauvinist," he isn't a chauvinist about any other attribute of terrestrial life, and in refusing to limit any further his ideas of what extraterrestrial life may be like he diverges from most biologists. Many of these think they can place further limits on the nature of extraterrestrial life on the basis of their knowledge of organic compounds. Most biologists believe that organic life anywhere must require nitrogen, for reasons that, though they cannot be stated precisely, as they can in the case of carbon or water, are nonetheless presumed to exist because nitrogen is found in all known amino acids and nucleotides and, therefore, by inference must be an essential ingredient of carbon-based life. If there was any nitrogen in the Martian atmosphere, many scientists thought at the time it would have to be less than 1 percent—the smallest amount that could be detected by an instrument aboard Mariner 9, which found none—and while some biologists thought less than 1 percent would be enough to support life, the outlook was less attractive than it might have been, in view of the fact that the Earth's atmosphere is 80 percent nitrogen.

Lederberg was not terribly concerned about the apparent lack of nitrogen in the atmosphere. He pointed out that we do not utilize the nitrogen in our atmosphere, but get it from plants, which take it from the soil. At the conference prior to the Viking 1 launch, he proposed that the reason Mariner 9 found no nitrogen was that Martian organisms had removed it from the atmosphere and combined it with the soil in the manner of legumes on Earth. But he made the proposal without much conviction, for although nitrogen may well have been present in the soil of Mars, no one had any way of knowing until Viking got there, and, in any case, not even Sagan had ventured to suggest that Mars was a huge compost heap. Nevertheless, Sagan wasn't concerned about the lack of evidence for nitrogen, either. "We find nitrogen in our own atmosphere, and we decide that it has to be good for life all over," he said—going on to place one epithet on top of another. "It's what I call the grandmother theory: 'What's good for her is good for me.' It's chauvinistic in the extreme."

Whereas most biologists believe that, when it comes to the basic ingredients of life, what is good for one's grandmother actually is good for oneself, there were a few who didn't dismiss Sagan out of hand. One was Klaus Biemann, the principal scientist for the GCMS, who said before the launch: "It's important to look at things in a nonterrestrial way, and that is harder for biologists to do than other scientists, because they have only one data point to work with—terrestrial life. Biochemists tend to be a little more open because their knowledge of chemistry lets them be a little more abstract about the possibilities. To me, it's inconceivable that on Mars things would be the same way they are here; I'd be willing to think of biochemistry without nitrogen, or even carbon, and most biologists wouldn't consider such a thing. So Sagan serves a valuable function by shaking them up a bit."

Shaken up or not, most biologists wished they had been able to ascertain the presence or absence of nitrogen in the Martian atmosphere before they began designing expensive instruments to look for life. As one of the Viking biologists put it, if they could have known definitely that there was no nitrogen in the atmosphere there, the instruments in the biology package would have been quite different. Another scientist said: "When you're exploring a planet, the first things you ask are what kinds of elements are there, what the temperatures are, whether there's an atmosphere, and what's in it—all these circumstances. Next, you address the organic situation. Then you might finally have the information you need to design a craft and instruments that could look for life. But we're not doing that. Imagine going to a spectroscopist with a black bottle and asking him to tell you what's inside without opening it—and without letting him know what the bottle, the host material, is made of! That's how we're going to Mars and how we expect to make our analysis. Clearly, it represents something other than logical scientific strategy." Sagan opposed this argument, saying, in a volley of negatives: "Since we know nothing about exobiology, we cannot determine what conditions would govern it—and therefore establishing ahead of time the conditions of the environment on Mars won't really help us to decide if life is there or help us to look for it." He unquestionably preferred to plunge into the unknown with a minimum of preconceptions.

Since Sagan's no-holds-barred approach resulted in such things as rigging cameras to catch tentacled squamous purple ovoids, the biologists' approach, if less imaginative, may have been the more constructive, at least with respect to Mars. Though some chemically exotic planets may exist in the universe, Mars did not appear to be one of them. And Earth analogy, whatever its limitations, was all the biologists, like the geologists, had to go on—another reason they were so closely guided by the evolution of life on Earth. If it was assumed that life on Mars had made it through the successive stages of chemical evolution to the point where the polymers of the genetic material and of proteins were formed, the next step in the development of life would have been for the proteins to wall off, by forming a semipermeable membrane, a small volume of the organic soup—water rich in ingredients for the formation of more genetic material and more proteins. The first cell would have been very primitive—it wouldn't have had a nucleus (something that evolved later). Nevertheless, on Earth, at least, as it multiplied it had devoured all polymers that might possibly have competed with it to become the progenitor of all later forms of life here.

Molecular biologists consider all life on Earth—an ameba, a blade of grass, a human being—to be descended from this original cell. This fact leads Sagan, contemplating the supposedly diverse riches of life in the universe, to call terrestrial life "provincial." Molecular biologists look upon the cell as such an inevitable development of organic chemistry that they hope someday to be able to synthesize one in a laboratory. They can make all the amino acids (there are only 20 in all terrestrial forms of life, an argument for the concept that all life derives from one cellular source), and they can make short nucleic acid chains without using components of living cells. But they have not yet synthesized proteins in the way nucleic acid does, without the use of parts of living cells, and the method of assembly of a cell's various constituents is still unknown. "We are still far from making a living cell, but we know what the problems are," Alexander Rich, of MIT, a member of the Viking biology team, said a while ago. "We can make some of the bricks, but we don't yet know how to put the bricks together to make the cathedral." Once the right bricks come together in the right way, the cell will "come alive"—that is, it will begin to reproduce itself and to evolve. Once biologists can make a cell in the laboratory, of course, they may be even more willing to conclude that the evolution of cellular life is a common phenomenon in the universe. And wherever in the universe the requisites for such evolution exist, the whole process, from the first primitive organic compounds to the first cell, may not take very long—perhaps as short a time as 50 million to 100 million years. (Curiously, the joining of several cells into multicellular organisms took many times as long.) On Earth, the oldest known fossils are single-celled organisms—bacteria and algae—that lived more than 3.5 billion years ago, just 1 billion years after the Earth was formed and 0.5 billion years after the Earth's molten period, during which no sort of evolution could have gone on. Since the fossil cells are all quite far advanced, a number of biologists now think they had to have developed rather quickly. The more quickly this can happen, of course, the greater are the chances that cellular life got started on Mars, where the proper conditions, if they existed at all, could not have existed for long. And if cellular life once got started on Mars, some biologists believed, it might have adapted to the increasingly harsher conditions, so that there could be living cells on Mars today.

Although the biologists believe that the cell, much as they know it on Earth, is the basic unit of carbon-based life everywhere, they are certain that its evolution would be different from one planet to another. To them, one of the most interesting things about the discovery of life on Mars would be to see what other pathways it took. "If we find life there, the first question we'd want to ask is: Is it related to terrestrial life?" Harold P. Klein, head of the biology team and director of life sciences at NASA's Ames Research Center in Mountain View, California, said. Although Klein lacks Sagan's freewheeling imagination, he likes to speculate, and he did so in a quick, forceful manner. "Is it made of the same proteins, fats, and carbohydrates that we are? The answer to that question leads to others: If it is related—that is, if it contains most of the same amino acids we possess on Earth—did our life and life on Mars arise spontaneously and result in the same general blueprint in both places or was there some sort of common origin, such as that implied by Lederberg's suggestion that organic compounds from space, as evidenced by those found on meteorites, constantly fell on the planets at the time of their formation? On the other hand, if life on Mars is totally unlike life on Earth—is

not based on the same amino acids or even on carbon—that
would be terribly important, for it would prove that life could
take multiple forms on the most basic levels. Then we would
want to know: What do the Martians have to replace carbon or
amino acids? Biology would come of age, and the study of life
would be a cosmic thing." He ticked off the alternatives as
though he were programming a computer to sift through a
number of models.

While Sagan agreed with Klein, he was less interested in
studying Martian cells in relation to terrestrial ones—with the
aim of learning more about life on Earth—than in seeing how
Martian cells might differ among themselves or combine to
form macrobes and, ultimately, in seeing how they bear on the
question of intelligent life in the universe.

Klein went on to say: "Alternatively, if it turns out that there
is no life or indeed no organic compounds on Mars at all, I
think there is a very good chance that we'd conclude that the
chemical evolution theory, which drove us to Mars, had to
undergo some reassessment. We would have to begin to wonder
how right we were in thinking there was life elsewhere in the
universe, for that notion is predicated on the idea that
chemicals quickly and easily get together to form life. The
theory has led us to expect at least organic compounds on Mars,
and if we find nothing we might begin to think that life is a
much rarer phenomenon than people now believe."

Sagan, of course, didn't agree with a word of this; if life didn't
turn up on Mars, he said, it would prove only that Mars was not
one of those planets where life could exist, and we would have
to look somewhere else.

A measure of the different approaches of Sagan and the
biology team was the odds they laid on the possibility of cellular
life on Mars. Klein believed that the odds were 1 in 50, and he
arrived at that figure in a methodical, computerlike manner: "I
start with the theory that life arises from chemical evolution,"

he said. "Now I have great confidence in that theory—I am, say, eighty percent sure of it. If it is true, what is the probability that advanced organic compounds arose on Mars? I'd give that a fifty-fifty probability. Then what's the probability that the organic chemistry reached a replicating system that still survives? I come up with two stumbling blocks to the continuance of life, if it got started. The first is the high incidence of ultraviolet light on the planet. Could life have developed a protection against this? I give that possibility one chance in twenty. The other stumbling block is water. It is hard to conceive of a biological system that doesn't include water, and any way of providing it on Mars involves hocus-pocus; perhaps there is permafrost underground; perhaps the microbes get their water from the soil. The water is the most serious question, and I give only one chance in ten of there being a way around it. An eighty percent chance divided by two divided by two again, and then divided by ten, is a two percent chance, or one chance in fifty."

Sagan, predictably, was reluctant to place any limit on the chance of Martian life, preferring to remain completely loose and uncommitted. In public, at least, he jauntily resorted to uncomputerish evasive tactics. "What do *you* say?" he asked a skeptical reporter who was pressing him for a statistical answer. "One in ten thousand? One in a hundred thousand? One in a million?" The reporter chose the last figure. "Okay, here's my penny," Sagan said. "Now where's your money?"

The Viking lander's three biology experiments, which collectively cost $55 million, were packed into a black box that measured 1 foot on each side and weighed (on Earth) just 20 pounds. They were a welter of golden pipes and hoses; round silvery cans containing chambers that rotated on carousels; and silvery and golden tanks of various gases, such as carbon dioxide (some of which was radioactive), helium, and krypton. Remotely controlled instruments could not simply look for life; they had to look for a particular aspect of life. In all, there were over 50 proposals for biology instruments of different sorts, including one by Lederberg involving a microscope.

To the extent that the scientists ever sat down and tried to figure out just what signs of life to look for on another planet, they decided that of all the indications they could reasonably examine, metabolism was the most foolproof, for in the course of that process all known forms of life, by means of respiration, photosynthesis, or other methods, react chemically with their environment. Whatever form life might assume, it would have to take in and discharge nutrients and gases in order to grow or to derive energy, and these interactions, going on inside sealed containers, would be relatively easy to detect. The three instruments were the pyrolytic-release experiment, of which Norman Horowitz, a geneticist and professor of biology at the California Institute of Technology, was the principal investigator or chief scientist; the label-release experiment, of which Gilbert V. Levin, of Biospherics, Inc., in Rockville, Maryland, was the principal investigator; and the gas-exchange experiment, whose principal investigator was Vance Oyama, of the Ames Research Center. Horowitz's pyrolytic-release instrument looked for the assimilation of atmospheric carbon—for example, the kind of metabolic activity that on Earth would be associated with plants: an autotrophic reaction such as photosynthesis. (A plant derives energy from sunlight, using it, in a reaction involving carbon dioxide, to get oxygen from water.) The two other instruments, Levin's label release and Oyama's gas exchange, both looked for a heterotrophic reaction, such as animals employ, deriving nourishment from organic compounds; Oyama's looked at a much broader range of possible reactions than Levin's, but was far less sensitive to these reactions. Whatever their differences, all three instruments involved the incubation of Martian soil in small airtight chambers and under varying circumstances, such as higher or lower temperatures, different amounts of water (including, in Horowitz's case, none at all), and (in the case of the other two instruments) varying quantities of liquid nutrient. And all three instruments monitored any subsequent changes of gases or soil content—two of them involving the tracing of radioactive carbon dioxide—which could be indicative of life.

Initially NASA had selected a fourth instrument to go along, which was not designed to look for metabolic gas changes, but it was later dropped. Called the light-scattering experiment, it was developed in the early 1960s by Wolf Vishniac—it was the first of all the biology instruments to be conceived and built. Vishniac, a short, stocky scientist with a close-cropped black beard, was the son of Roman Vishniac, the microbiologist and photographer; at the time, he was on the faculty of the Yale Medical School but soon moved to the University of Rochester as professor of biology. Vishniac, who would later propose the neo-Lowellian theory that the seasonal markings on Mars were caused by dust landing on plants and falling off again, was a prime mover, in the early 1960s, in the inception of the landings on Mars, and in particular in the inclusion of biological instruments on the spacecraft. Ten years older than Sagan, Vishniac also had been interested in life on Mars since he was a child. He was like Sagan in many respects, thinking nothing of dashing off a three- or four-page letter about any minor point that displeased him. He, too, had lectured the Apollo astronauts about what signs of life they should look for on the Moon. He was present at a 1959 meeting of the Space Science Board of the National Academy of Sciences, of which he, like Sagan and Lederberg, was a member, when the possibility of sending life-detection instruments to Mars was already under discussion; at the meeting, Thomas Gold, the astrophysicist who would later bring Sagan to Cornell and who was as forthright then as he is now, asserted that no biologist had yet developed an automated instrument that was capable of detecting life remotely on another planet. Vishniac, stunned by the veracity of this pronouncement, immediately set to work trying to build such a device. Two years later, in 1961, he completed the instrument, which not unnaturally came to be called the Wolf Trap; it sucked in dirt, bubbled it through water, and then shined a light through the solution, on the theory that over a period of days the light coming through, which was being measured, would get dimmer if any organisms inside were replicating. The Wolf Trap was the forerunner of Vishniac's ill-fated light-scattering experiment for Viking, which worked the same way.

In the early 1960s other men were at work on the prototypes of instruments to detect life on Mars. Among them was Gilbert Levin, then a sanitation engineer who was responsible for monitoring pollution in the water supply in towns along part of the California coast and who had invented an automated method of detecting microbes by means of their metabolic release of carbon, which he thought might be useful on Mars. At that time, the clearinghouse for ideas for life-detection instruments on Mars was the office at J.P.L. of Gerald A. Soffen, who would later be Viking project scientist and chairman of Viking's science steering group. Soffen introduced Levin, with his machine for detecting microbes in water supplies, to Norman Horowitz at Caltech, which was just across Pasadena from J.P.L., and the two of them refined the instrument. They called it Gulliver because it was meant to journey to far-off places and encounter strange forms of life—something it initially was designed to accomplish by ejecting a sticky string onto Martian soil and rolling it back in again so that the soil, and any organisms in it, could be incubated.

All these ideas were developed not for Viking but for an

earlier Martian craft, Voyager, which had been conceived in 1963. Voyager, which was far more ambitious than Viking, involved two landers that would have been launched by Saturn rockets; they would have weighed a total of 50,000 pounds and cost about $4 billion. "It was a very sophisticated biological laboratory, which Horowitz, and also Lederberg and all the other Nobel laureates, would have operated directly from the ground, pushing buttons and pulling levers for their experiments at J.P.L.," said one geologist, who liked the biologists' grandiose schemes then even less than he did their more moderate ones now. "President Johnson killed it in December of 1967, and thank God."

The only person who regretted the demise of Voyager (the name later would be given to the two highly successful spacecraft that would fly by Jupiter, Saturn, and other planets beyond) was Sagan, who described himself at the time of Viking as an "unreconstructed Voyager enthusiast." Viking, a much more modest program, was approved in 1968. As Viking's headquarters, under James S. Martin, Jr., the project manager, would be at the Langley Space Center in Hampton Roads, Virginia, Soffen, the project scientist, moved to Langley. During most of the next 8 years, when the plans for Viking were being made, the 13 team leaders, for both the orbiter and the lander, reported to him there. (There were about 70 scientists on all the teams, though the number swelled during the mission.) At that time, the leader of the biology team was Vishniac, and the other principal members, then and now, were Klein, Lederberg, Rich, Horowitz, Oyama, and Levin. With their associates, the team would eventually number about twenty.

The seven principal members were a contentious lot—most of them, and a few of the associates as well, were highly individual scientists, more used to running their own laboratories than to cooperating in joint ventures—and they were unable to agree on the best way to look for life on Mars. "When the time came to start getting things done, Vishniac was not able to keep this menagerie on the track," one team member said later. "Vishniac was basically a very gentle person. At meetings, he let everyone have his say—but at the end, there was never any sense of forward movement." Klein, who had gone to Brooklyn College with Vishniac, replaced him as team leader; he had a quick, bright, authoritative way of talking that could, and would, lay down the law under far tougher circumstances. A thickset man with penetrating sharp eyes and a continually quizzical look, Klein had been chairman of the department of biology at Brandeis University in Boston before joining NASA. Vishniac, of course, remained a member of the team, as the scientist in charge of the light-scattering experiment. A year later, though, in 1973, the engineers determined that, for reasons of weight and complexity, one of the four instruments had to be dropped. A subcommittee that included Klein, Lederberg, and Rich, who were senior members of the team but had no experiments aboard Viking, determined that Vishniac's had to be the one to go for it depended on the greatest number of prior assumptions about life on Mars—not only that it would survive in water (a dubious proposition)—but further that it would replicate there.

The three biologists with instruments aboard Viking differed among themselves almost as much as each differed from Carl Sagan. Though Sagan was at the center of many of Viking's disputes, other storms raged all around him, interlocking in a giant cyclone system. Viking, with all the issues it raised about life on Earth as well as in space, couldn't help being a tumultuous affair. One argument was about the amount of water used in these experiments. Horowitz, arguing that life on Mars would be accustomed to extremely dry conditions, said that Oyama, who (in Vishniac's absence) used the richest nutrient and the most water, was looking for Earthlike life on Mars.

Oyama maintained, however, that the more the Martian microbes were plied with food and drink the more gases they would generate. "A man doesn't respire much in bed, and Martian microbes won't respire much, either, unless we perturb them," he said. "The more they get of what they want the more there will be to measure." Horowitz retorted by christening the instrument Oyama's Pharmacy.

Horowitz had come to have a very low opinion of the chances of finding life on Mars at all: he rated it as "low—not quite zero, but very low." Like Sagan, Horowitz—a pragmatic man who had been born in Pittsburgh in 1915—had read Burroughs's books when he was a boy, but unlike Sagan his imagination about life in space had not been fired up. What got him thinking along those lines was the works of Darwin, which he read in high school. He had been at Caltech, where he got his doctorate in 1939, for most of his professional life. From 1965 to 1970, he had the additional chore of heading the biological section of J.P.L. He made the major discovery that one gene governed one enzyme in the early 1950s; later he came to feel that the discovery was a powerful tool in studying biochemical evolution, and this was what first interested him in the search for primitive life on Mars. In the late 1950s and early 1960s, he joined a discussion group organized by Joshua Lederberg, who was then at Stanford, and which included Sagan when he was at Berkeley and Stanford, as well as scientists at the California Institute of Technology. The group, called Westex, just like a brother organization at the Massachusetts Institute of Technology called, naturally enough, Eastex, which included Biemann, Rich, the physicist Philip Morrison, and others, talked about such matters as the best ways to look for life on the other planets and the need for a quarantine. However, in 1965, when Horowitz was still associated with Levin on Gulliver, Mariner 4 made the discovery that the pressure on Mars was not sufficient for liquid water to exist on the surface of the planet. This news affected Horowitz more profoundly than it did most of his colleagues; he became convinced that even Levin's instrument, although it used less water than Oyama's, would also drown the very organisms it was looking for. Since he felt that neither Levin nor Oyama was being what he called "tough enough on the evidence," he and a colleague, George Hobby, began work on a different sort of instrument—in a way, it was Levin's machine turned upside down—that required no liquids or nutrients.

Horowitz's pessimism was rooted in part in his conviction that no terrestrial life could live and grow—that is, exist in other than a dormant state—in the Antarctic valleys, the coldest, driest, most Martian spots on Earth. The Earth, of course, is otherwise a planet with a rich and abundant corpus of life continually pushing into all sorts of exotic ecological niches. If, despite this biological vigor, no form of life had developed on Earth that could make its home in the Antarctic valleys—that is, live actively to include replicating—what chance was there that life did the same things on Mars, which was generally far colder and drier than those coldest, driest places on Earth? Though this argument was more suggestive than conclusive, Horowitz found the suggestion compelling. Vishniac, whose zeal about life on Mars was greater, if anything, than Sagan's, and whose own experiment—far from taking the dryness of Mars into account—had been wetter even than Oyama's, emphatically did not. He didn't even accept Horowitz's proposition that the cold dry valleys were devoid of active forms of life. He went to Antarctica; as his own instrument had been eliminated, he took Levin's with him. One day, when he was working alone, taking core samples of the soil, he slipped and fell to his death. Sagan, who later wrote his obituary for *Icarus*, sees Vishniac as the second man in history to die for the idea of extraterrestrial life—the first having been Giordana Bruno, who in 1600 was burned at the stake for believing in, among other things, a multiplicity of worlds. The question of whether organisms not only exist but replicate on the surface of the Antarctic valleys is still unresolved.

The biologists were a beleaguered group, for not only were they forever bickering with one another, but they also had to fend off the criticism of the geologists, who tended to think they were all watered-down versions of Sagan. Just as the geologists feared Sagan might mistake a rock in one of the photographs for a Martian macrobe, they feared the biologists would believe themselves to have detected microbes with their instruments when there were none—a scenario one geologist described as a "horror show." The biologists felt that enough checks had been built into their instruments so that this wouldn't happen; all three of the experiments had control procedures. The geologists often worried about another problem. What if the different scientists' instruments gave conflicting readings? For example, the three biology experiments might all register the presence of life, but Biemann's GCMS might not find any trace whatever of organic material in the soil. "In that case, I'd tend to put my faith in the organic chemistry experiment and wonder what went wrong with ours," said Horowitz, who was as skeptical about the life-detection instruments as he was about life on Mars.

Even more fundamental than interpreting conflicting results from all three experiments was the problem of how to interpret what was a positive indication from just one experiment. Because of delays in the completion of the instruments and because of a number of cost overruns superimposed upon a general picture of NASA budget cuts (the biology package, and also the GCMS, cost three times their initially estimated costs of $18 million each), the biologists, at the time the instruments were launched, had had neither the time nor the money left over to calibrate their instruments. Three months before the launch of Viking 1, Klein said: "What *is* a measurement consistent with life on Levin's instrument—a reading of, say, two hundred or four hundred or six hundred? Originally, we were going to use the hell out of these instruments on different materials on Earth, such as desert soils, Arctic soils, rich Californian soils, and soils known to be sterile, but about a year ago, in a cost reduction, we were told there was no money to operate the laboratory test models of the instruments. What NASA was saying was that here's some expensive instruments, but there's no money to build up a baseline for data—in other words, no money to learn to use them. NASA, of course, has its own budgetary problems. But if we do not get many months to test and find out what is a significant reading, it will be very embarrassing, because we might get results and then, in front of the whole world, not be able to interpret them." Though Klein finally *did* get the calibrating tests he wanted, he had to scrape together the funding, and even then the tests were not done until the last possible moment—while the two spacecraft were well on their way to Mars.

Many Viking scientists, far from wondering how the biologists would interpret their data, seriously questioned whether the biology package would work at all. Three months before the launch of Viking 1, the instruments, which had all been tested individually, had not yet had a completely successful test of the integrated biology package—that is, a complete run-through of all of them together rehearsing every step they would perform on Mars. In December 1973, when the first integrated test of the biology package was attempted, all the gases—the radioactive carbon dioxide and so forth—flowed through the pipes and in and out of the incubation chambers as they should have, but as soon as the soil was introduced, there was what Klein, whose face is capable of assuming a wonderfully mournful expression, termed a "catastrophe." The carousels, which rotated the incubation chambers under the hoppers for the soil, got stuck. Certain pieces of metal got bent out of shape. Some of the detecting devices that looked for the sort of gas changes that might indicate life (those for radioactive carbon dioxide in particular) went out of scale. "The tests showed that we did not have viable instruments," Klein said. The biology team, as well as engineers at both NASA and the prime contractor for the package, TRW Systems Group, went into a crash program, in the course of which they identified some 80 problems, including the fact that the detectors for the carbon dioxide were sensitive to light and heat as well.

By January 1975, 7 months before the launch, there was a whole new version of the instruments. The integrated test, which was to last 2 months, began in February 1975. The new package—one of several duplicates, as the ones bound for Mars were already at Cape Canaveral—was put inside a vacuum chamber that simulated the Martian environment, not only its low pressure but its low temperature as well. At first almost everything worked beautifully—particularly the flow of gases, the instruments' heaters, valves, and detectors. However, when the soil was dumped in and incubations started, the only instrument that gave any usable data was Oyama's. The other two didn't work. After 2 weeks, those two, Horowitz's and Levin's, were removed from the vacuum chamber. Levin's turned out to have a gas line plugged with solder, and two plungers that were supposed to break open two ampules of liquid nutrients had failed so that the nutrients never reached the soil. Both instruments were repaired and put back in the vacuum chamber to continue the tests. Again, there were no results from Horowitz's and Levin's instruments, though Oyama's continued to give creditable data. At the end of March, when the test period was over, and with 4 months to go until launch, the instruments were removed from the chamber and examined. "When we opened them, we could see soil spewed all over the insides: there had been an explosion," Klein said. "We went into a crash program to see what had happened, and now we think we know: when we had taken the two experiments out of the vacuum chamber after the second week, we had assumed that they were tightly sealed, but they weren't. When they were outside the vacuum chamber, at room temperature and pressure, the pressure inside the instrument rose from the five millibars of Mars to the one thousand millibars of the Earth. Then, when we put the instruments back inside the chamber, at five millibars, they blew up! Now, we can say that such a thing would never happen on Mars, but at the same time we have not yet had a successful integrated test with soil in the instruments." In the remaining weeks before launch, the biology team was able to rerun the test, but for a shorter time than necessary; though it was not a complete success, Klein said later that the results were "a bare minimum to give us a reasonable confidence."

And the troubles were not limited to the biology instruments alone, but extended to the GCMS as well. Five months before the launch, with the instruments already aboard the spacecraft, Biemann, using a duplicate instrument at the Ames Research Center, discovered a problem: when he ran some test soil through the duplicate, he found traces of organic chemicals when there weren't any. A major flap ensued. It turned out that the fault was with the grinder that broke up the soil. The grinder was made of steel; there was carbon in the steel; and as the grinder ground some of *itself* into the soil, the carbon combined with the hydrogen that was used to flush out the instrument between experiments to make a hydrocarbon—an organic chemical—that the GCMS proceeded to identify. "The grinder for the GCMS put steel filings in the soil!" one geologist said, both appalled and gleeful. "Hell, anybody knows that will happen!" This was basically an engineering problem, for which no one blamed Biemann, whose GCMS was now otherwise in considerably better shape than the biology instruments. "What screws us up isn't the complex scientific problems, but the high school problems like the grinder," another scientist said. Before launch, Biemann managed to get the hydrogen flush gas replaced with another that wouldn't combine with carbon in the grinder to give a false reading on Mars.

"They've spent fifty-five million dollars or sixty million dollars on those instruments, and they're still not sure they will work!" one geologist cackled. The inadequacies of the biology instruments in particular drove the geologists to the sort of vituperation they normally reserved for Sagan. "Thank God Viking wasn't launched three years ago, when it was originally scheduled, for if it had gone in 1973 they would have flown a pile of junk," a second geologist said. Being old hands in space, the geologists made no secret of the fact that they thought of the biologists as duffers outside their own planet and maybe on it as well. The geologists, who had 15 years experience designing instruments for use on other planets and dealing with aerospace contractors who sometimes promised more than they could deliver, felt that the biologists should have called on them for advice. The biologists thought the geologists were jealous because of their lower priority on Viking. Klein pointed out that the biology instruments were much more complicated than any instruments ever flown in space before, including the geologists'; this was the first time life-detection instruments had ever been sent into space.

The very novelty was what the geologists regarded as the source of the trouble, however. They saw the difficulties with the biology package as just one more reason the exploration of Mars should have been done in steps; the instruments, they said, might have been less expensive and more reliable if the biologists had had a better idea of what they were doing—a statement most biologists, including Klein, agreed with. One geologist said: "With the biology instruments, we're going to

Mars with too much too soon! Not only are we failing to go to Mars in a logical, stepwise fashion—where we learn first about the atmosphere, then about the chemical composition of the planet, before designing instruments to look for life—but in order to leapfrog ahead, the biologists are using radically new instruments, the simplest versions of which have never been flown in space before! We could have sent atmospheric probes or hard landers with chemical instruments to Mars for the cost of the overruns on the biology instruments alone!"

At bottom, most scientists didn't blame the biologists for the troubles, but rather the NASA system, in which engineers, not scientists, were in charge of what the scientists saw as scientific decisions. Indeed, it was engineers at NASA headquarters—as well as politicians in Congress—who finally decided that Viking would include a biology package. To the engineers and the politicians—as well as to Sagan—life was clearly "the most pizzazzing idea around." Many biologists, feeling they were being asked to hold some sort of bag, maintained that finding life on Mars had never been Viking's chief purpose. They pointed out that they were just one of Viking's 13 scientific teams and that the 1 cubic foot allotted to their package was no more than the space given to other instruments. If matters had been otherwise, they would have had a lot more space. They felt, though, that in recent years NASA had highlighted the search for life "almost certainly with the intent to keep Viking alive by tweaking the public interest with the possibility of life on Mars." Almost everyone connected with Viking believed that there would have been no exploration of Mars if it weren't for the search for life. As one biologist said, possibly getting a little of the biologists' own back: "You don't seriously think Congress would have spent one billion dollars just to do *geology* on Mars?"

The biologists, of course, were not about to stop this manna from heaven, though if they had had more say in the matter, they too would have preferred to send craft to Mars in a slower, more deliberate fashion. However, the careful, step-by-step program that most scientists would have preferred never showed much sign of materializing. It was true that Viking itself was supposed to have been the first of a series of landers, and not a one-shot culmination of the exploration of Mars, which was what it was now being considered by NASA. Sagan and Lederberg, who usually thought on a grander scale than most other scientists, had always tended to regard Viking itself, expensive as it was, as a preliminary probe—the first in an ever-increasing armada that might culminate someday in a manned landing.

In Florida before the launch of Viking 1, Lederberg—possibly adopting the cautious manner of one who realized a day of reckoning was not far off—was suddenly quite pessimistic about the chances of life on Mars. No unexpected compounds or abundances of the sort that betray life, he said, had been found either on Mars or in its atmosphere. If there were Martian scientists, they wouldn't have to leave their planet in order to detect the presence of life on Earth, for it would show up remotely in (among other things) the presence of both oxygen and methane in our atmosphere (another example of thermodynamic disequilibrium). Methane, being chemically incompatible with oxygen, should probably have disappeared by now; it hasn't, because life continues to manufacture it. It is produced by bacteria that live in marshes and in the intestines of cows and other grass eaters. "There's no phenomenon observed so far on Mars that says there has to be life there," Lederberg said, "and that's dismal."

As Sagan and Lederberg bobbed around in the pool at the Ramada Inn at Cocoa Beach, Florida, where they were staying before the launch, they talked about what the effect would be on the odds for the existence of life elsewhere in the universe if the biology instruments gave a unanimous negative to the question of life on Mars. The matter wouldn't come up, they decided, because the instruments aboard Viking—assuming they worked—were incapable of providing an irrefutable negative. Sagan's fertile imagination had churned out a number of scenarios showing how the instruments might fail to detect life even if it was present—thereby standing on its head the geologists' argument that the instruments might register life that was not there. For example, he said, the spacecraft might land in a part of Mars where there was no life; that is, it might fail to land in one of Sagan's and Lederberg's low, warm, wet microenvironments. Even if one assumed that the spacecraft landed in a Martian Garden of Eden, and the instruments were working properly, there were plenty of holes Martian organisms might slip through undetected. "Take the nutrients in Oyama's experiment," Sagan said. "We're sending food to the Martians, but what we're sending may be poison to them. Many simple organic compounds are poisons for us, such as hydrogen cyanide. So the food we're sending them might kill them. Or say the food doesn't kill them; rather, they don't like it because they eat different stuff. So the bugs inside the instrument don't eat it and can't be detected—but meanwhile there are other microbes outside, chomping away on the spacecraft's zirconium paint, say, which happens to be their bag. And of course, the microbes might love the food we send them and eat it all up, but instead of oxidizing it into carbon dioxide they might only ferment it to butyric acid, which can't be detected either by Levin's instrument or by Oyama's. Then we get no reading— and we say, 'There's nobody eating our food.' And certain kinds of life could be missed by Horowitz's experiment. This is not a criticism of the instruments; it's merely to suggest the complexity of biology. Looking for life is hard."

Lederberg agreed with Sagan that Viking probably couldn't

provide a definite negative and that even if it could the result would have little bearing on our ideas about the existence of life elsewhere. Mars might be a planetary red herring. It could well be that the solar system as a whole is too arid and inhospitable to support life on more than one planet, and certainly conditions on Mars are so harsh that the absence of cellular life there would not prove much. Viking *was* capable though of providing a positive answer—on that point Sagan and Lederberg were agreed. And because of the harshness of the environment, the existence of life on Mars would speak volumes about an abundance of life, including intelligent life, all over the universe. Sagan, treading water, looked pleased with this heads-I-win-tails-you-lose proposition.

It took 11 months for the spacecraft to arrive in the vicinity of Mars. As Viking 1 approached the planet, all the scientists, whatever their stripe, were suddenly becoming very solicitous about its success; big, controversial projects have a way of generating loyalty at the last minute. (Similarly, last-minute preparations, such as the scientists had been going through for the last year or more, are often accompanied by a good deal of grousing.) Now, briefly, there was a period of relaxation. All hands agreed that if any craft were to make a soft landing on Mars, it would have to be something on the order of Viking or even more complicated—and it would obviously include biology instruments. "It's a gutsy thing," one particularly critical scientist said. "The only question now is: Will it work?"

Part Two

KLEIN, HOROWITZ, LEVIN, AND OYAMA

Important, Unique, and Exciting Things

Ideas, like all living things or their products, evolve; science in particular is in unceasing conflict not only between ideas but between the scientists who espouse them. The search for life on Mars has had its ups and downs since at least the end of the last century, and the arrival of the first Viking spacecraft, Viking 1, which went into orbit around Mars on June 19, merely sped up the process. Indeed, while Viking 1 was still far out in space, all its cameras could see of the planet were the tips of the four huge Tharsis volcanoes sticking up above layers of haze—just as had been the case 5 years before, when Mariner 9 had approached Mars, only that time the planet had been engulfed in a global dust storm instead of clouds. Some of the Viking scientists at the Jet Propulsion Laboratory were reminded of a movie that had recently opened, the remake of the film *King Kong*, in which explorers searching for an island suspected of harboring odd forms of life found it shrouded in a cloud bank. The Martian haze proved to be very light, and as Viking got closer the

113

cameras could easily see through it; a water detector aboard the orbiter (there had been no such instrument aboard Mariner) determined the gossamerlike clouds to be made in part of water ice crystals. Though this was not the first time water had been identified on Mars, the chances for life shot up. Then, when the lander left the orbiter and descended to the ground, which it did about a month after going into orbit, an instrument on the aeroshell—the outer casing that protected the lander from heating in the atmosphere—found that the rare Martian air was 2 or 3 percent nitrogen. (Clearly the earlier finding that Mars's atmosphere could be no more than 1 percent nitrogen was in error.) The chances for life soared. All the elements necessary for life were now known to be present on Mars—though whether they had actually *formed* organic compounds, let alone living things, remained to be seen.

This news was followed by a setback, for the amount of argon in the atmosphere turned out to be only 1 or 2 percent, far less than the scientists had been led to expect from the Soviet information. (The Martian atmosphere was 95 percent carbon dioxide; the remaining 5 percent, in addition to nitrogen and argon, included carbon monoxide, oxygen, and trace amounts of other gases.) Further analysis of the argon content suggested that at one time the Martian atmosphere could have been perhaps 100 times as dense as it is now, confirming that there might have been sufficient pressure to allow small bodies of standing water to exist and for channels to flow, but in the opinion of most scientists, not enough for any large bodies of water, such as oceans or even lakes, to have existed for any period of time. The chances for life fluctuated, for opinions differed about how much water was needed for the origins of life. All in all, Klein, the biology team leader, was content to let his odds remain at his previous 1 in 50.

As Viking's orbital cameras were able to take a great many more high-resolution pictures consecutively than Mariner's, the scientists saw far more of Mars in detail than before; they observed, for example, many more channels with tributary systems, suggesting rainfall, than they had earlier—and not just around the equator, as Mariner 9 had observed, but all over the planet. (The odds went up a notch.) When Masursky, the leader of the site certification team, received the first Viking photos of the prime landing site—at the mouth of the four great channels debouching onto the southern rim of the Chryse basin—he felt as though he had put on a new pair of glasses, and what he saw alarmed him. Where the Mariner photographs had revealed what looked like a safe sort of alluvial area, Viking discovered that the great channels were still deeply etched into the ground; the four streambeds did not finally run out onto the smooth plain for another 100 kilometers or so to the north. The landing had to be postponed. "Nobody thought a crash landing on July 4 would be a good Bicentennial present," Sagan said.

First, the landing site was moved 100 kilometers downstream out into the Chryse plain and beyond the true river mouth area, but it did not remain there very long. As the Viking orbiter's imagery proceeded swath by swath, day by day, to the north, it revealed other types of unsettling terrain downstream—in particular, the ground far out onto the plain was littered with large boulders, ejecta from craters, perhaps, but more likely, in Masursky's view, the flotsam and jetsam from ancient floods that had carried the boulders downstream from the highlands. The landing site was pushed another 300 kilometers north, out into the center of the plain, which looked relatively block-free. But there was trouble there, too, for data from the radar antenna at Arecibo, Puerto Rico—operated by colleagues of Sagan's at Cornell—was, for the first time since the landing site had been selected in 1974, able to receive signals bounced from the Chryse region. The signals from the center of the basin, where Masursky now wanted to land, were low in reflectivity, indicating (in the opinion of the Cornell scientists, at least) that the area was too soft for a safe landing, possibly because of deep dust. Sagan and his colleagues (who, though they were not on Masursky's site certification board, acted as consultants) argued for a landing site higher up on the rim of the Chryse basin, where the ground appeared stronger. Masursky, who didn't believe in deep dust, pointed out that the rim was where the big boulders were. In the end, there was a compromise. Viking 1 would land halfway up the basin slope, where neither blockiness nor softness (if that is what it was) would be extreme. The spot, midway up the western slope, was just below the mouths of two other big ancient streambeds; they appeared to have arisen in the collapsed highland terrain just west of Chryse, where very possibly they had had their sources in vast deposits of underground ice that had been melted very quickly by volcanic activity. All around, the landscape showed signs of the rush of ancient floodwaters—craters hundreds of kilometers below the channel mouths had apparently once been filled with water, which then broke through on their downhill sides—a giant flood, 300 kilometers across.

The Viking lander touched down 2 weeks late on July 20, when it was 5:12 A.M. at the Jet Propulsion Laboratory, and late afternoon in the Chryse basin, though no one on the ground could confirm the fact for the 19 minutes it now took—because of the delay in landing—for a signal from Mars, more than 206 million miles away and nearing the most distant point from Earth on its orbit, to reach J.P.L. Mutch, leader of the lander imaging team, delivered a commentary on the first couple of photographs as they appeared, line by line, on the television monitor. He was nervous. Although he had thought a great deal over the last 8 years about the scene he was shortly to see, and although he and his fellow scientists had made a number of guesses concerning what it would look like, he knew there might be unexpected elements, and he was worried at the prospect of looking at a scene he didn't understand and having to go out on a limb before millions of people. "There's the first piece of information coming in!" he said. "Yes, yes, that's it. Rocks! That's beautiful!" Mutch was silent for a while, in part because there wasn't anything he could say about the rocks and in part because he was overcome, not only by the fact that he was seeing the landscape of Mars from close up for the first time, but that he was seeing it so clearly. Over the years, a number of Mutch's colleagues had thought he was going out on a limb just spending 8 years working on a camera to go to Mars. "They thought I was daft!" he said a little later. "And now they see this! There's nothing like working eight years on something and then have it come across. It's like mountain climbing— trying to climb Mount Everest: there's a slim chance you'll make it, but if you do, success is unequivocal!"

Some of the rocks were angular. There was a good deal of dust between them. Most of the rocks appeared to Mutch to be complex; some of them were angular and faceted, leading him to suggest that on Earth, at least, they might have been sandblasted by windblown dust. Some of the rocks had dust piled up on their lee side, the way rocks on deserts here do. The signs of wind, Mutch said, were one way he could tell at a glance that Viking had not landed on the Moon; another way was that the Martian atmosphere—rarefied as it was—softened the shadows; on the Moon, where there is no air to scatter sunlight, the shadows are hard, precise, and largely impenetrable. "Isn't that the footpad coming into view?" Mutch said. Indeed it was—contrary to certain expectations, it had not sunk into the ground but rather was solidly on top. The footpad, which was concave, was filled with dust blown there at the time of the landing. Mutch pointed out the rivet that held the footpad together as an indication that the photography was remarkably clear. Sagan, who was down in the press center being interviewed by one of the networks, said later he had never in his life been so glad to see a rivet.

Though Sagan might have made more of the first photographs from Mars, Mutch's presentation was more spontaneous. Now that it was becoming apparent there would be nothing to cause him to go far out on a limb, and possibly saw himself off, he began talking so enthusiastically that he lost his voice and didn't regain it for 3 days. Three minutes after the first picture was complete, the second picture, a panorama toward the horizon, began to come down and assemble on the screen. "We can see the horizon! Oh, isn't that sensational!" Mutch said. "There's a very bright sky! Oh, isn't that lovely! That's fantastic. That's just lovely." The loveliness was in fact more in the eye of the beholder; the view was like a nondescript, flat bit of desert—indeed a contest at J.P.L. for the photograph that would turn out to look most like Mars was won by a snapshot taken in a particularly arid section of the Mojave Desert, 50 miles away. Much of the Viking scientists' enthusiasm was, of course, for the successful culmination of a mission they had worked on for 8 years—as well as for the first glimpse of the surface of Mars. Mutch, over his microphone, could hear a cheer from scientists in another room; it was the members of the meteorology team, whose instrument was supposed to have flopped out from the side of the lander a few seconds after touchdown, but who had no way of knowing whether this had happened until they saw the instrument—a white box on the end of a white pole—form on the screen. The temperature at Chryse in summer turned out to range from about $-122°$ Fahrenheit in the early morning to a high of $-22°$ Fahrenheit in the early afternoon, the wind hovered around 15 miles per hour, and the pressure was 7.70 millibars and falling. (Another instrument that was supposed to come into operation soon after touchdown, the seismometer, did not start up and was the only one of the Viking experiments that never worked.) As the camera panned to the right, other elements of the spacecraft came into view—the housing for the sampler arm, which wouldn't go into operation until the eighth day when it gathered soil for three of the instruments, and the instruments' three hoppers that the sampler arm would drop the soil into. Funnels—in sections like petals—sprang up around them. Further on was the streamlined-looking shielding for a fuel tank. As the mission progressed, the recurring views of the spacecraft's superstructure would be pleasing to Mutch and the other scientists, who were always glad to see something familiar in the otherwise alien landscape.

As the camera continued swiveling to the right, the picture became longer and thinner—very much as though the scientists were peering through a narrow slit in a gun turret, scanning the horizon for signs of danger. Danger there was aplenty, for the landscape was littered with boulders high enough—18 inches or more—to have punctured Viking's underside had it landed on one. Far to the right, the wind had piled up dunes like ocean waves. Over the coming weeks and months, Mutch would be continually amazed at how much variety the cameras managed to pick out of what was essentially the same picture, simply by zeroing in here or there at different times of day; there seemed to be an infinite number of views in the same panorama. A few of them, he thought, were worthy to be called art. He was particularly taken by a high-resolution photograph of the sand dunes; it was shot in the evening before sunset, so that it was backlit. "It's the sort of picture a professional would take, but an amateur would not," he said. "It has a form to it. As a work of art, it's striking; people respond to it, perhaps because it relates to our own experience on Earth. It is a real Ansel Adams picture." And Sagan said, a little later; "You can't look at that picture—the gently rolling horizon, these drifts like a terrestrial desert, the ray of stones and boulders—without wishing you were in it."

The first pictures had been in black and white, for reasons of speed in transmission. The color photographs came in a few days later, and when they did, it turned out that the ground—the rocks and the dust—was uniformly orange red. The most likely explanation was that the rocks were coated, probably quite thickly, with iron oxides such as hematite, ordinary rust; presumably, the iron oxides were fixed onto rocks with the aid of ultraviolet light from the Sun. (This is not possible on Earth, whose atmosphere is much thicker and whose upper layers contain ozone, which filters out the Sun's ultraviolet light. Nor is it possible on the Moon, which has no atmosphere and hence no ambient oxygen to make the iron oxides in the first place.) The rusty redness of the landscape was not unexpected since Mars appears red from the Earth even with the naked eye; what was unexpected was that the sky, too, was red. This was not apparent in the first color photographs, transmitted in a series of dots called pixils, that were released with a blue sky. However, when James Pollack went over the pictures with a computer, pixil by pixil, he discovered that there were 25 percent too many blue ones; when the blue was lowered by that amount, the sky turned red. Preconceived ideas may have played a part in the error, and not for the first time in the study of Mars: most scientists had thought the sky would be blue, at least near the horizon where the atmosphere was thicker, and veering toward the black of space overhead. Now, however, they kicked themselves for not having guessed that some of the red dust particles, as small as lint, would remain suspended in the air between dust storms, which occur mainly in the Martian winter, so that the sky would be permanently red; though the sky might brighten as the particles fell out, enough would remain so that even the sky directly overhead, where the atmosphere was thinnest, would never be black and starry in daytime as on the Moon. Paradoxically, the Sun itself, so red in our own sunsets, actually looks blue in Martian ones because the dust particles are more effective at bending blue light at small angles than red light.

At first, of course, everyone had kept an eye out for Sagan's macrobes; no one wanted to be caught unprepared if by some chance a silicon giraffe or a squamous purple ovoid should wander by. Macrobes seemed to be a highly charged subject. Many scientists seemed nervous in the self-conscious, self-mocking way of people who hold a ticket to a lottery which they have a one-in-a-million chance of winning. Cartoons by Viking staffers sprinkled the J.P.L bulletin boards—a picture of a Martian, with three legs and antennas just like the Viking lander, sidling up to it with obviously amorous intentions; another drawing of Viking at Chryse, a vine growing round and round it.

Whatever people's hopes, there was no visible sign of life in the barren landscape (which no one even now could call barren with absolute certainty). "There was not a hint of life—no bushes, no trees, no cactus, no giraffes, antelopes, or rabbits; no burrows, tracks, footprints, or spoor; no patches of a color that might be photosynthetic pigments," said Sagan, who by and large took well what many regard as a defeat for him—although, strictly speaking, he had simply been holding the door open to these possibilities, his colleagues felt he would not be happy if the door were shut. The situation was, of course, hardly unexpected, even by him, and he was quick to concede the obvious. He, and others as well, soon lowered their sights to searching the rocks for any uncharacteristic shapes, though this, too, was fruitless. "There was not a single recognizable funny-looking thing, no obvious sign of thermodynamic disequilibrium," Sagan said. Some of the biologists showed a short-lived interest in oddly shaped rocks, such as a cylindrical one that had been named the Midas Muffler and a concave one known as the Dutch Shoe. Sagan was attracted briefly by a suspiciously spherical rock in the foreground. Mutch thought certain biologists appeared almost to stake claims on certain rocks, as though they wanted to make sure they got the credit should it prove to be alive. Though Sagan continued to search the photographs methodically for signs of life, before long he conceded: "So far, no rock has obviously got up and moved away."

Some of Sagan's colleagues thought he seemed a little quiet and subdued during the mission; aside from a few appearances on television, he kept what was, for him, a low profile—an impression that was accentuated by his having an office on the fifth floor of the tall, glassy headquarters building at J.P.L., removed from the offices of the other Viking scientists in a bronze-colored building erected especially for the project. Sagan didn't even object when the imaging team decided to cancel the single-line scans—those occasional up-and-down nods of the cameras' diodes to detect motion, for which he had fought so hard. (The scans would be resumed in April, when the high winds began, to detect movement of dust.) Sagan was in an anomalous position in more ways than one. In the absence of anything demonstrably macrobial, the imaging team's data were chiefly of use to geologists, not astronomers or biologists. And the data from the biology instruments, in the first instance at least, would be the preserve of the biology team. Sagan would, of course, study the manner in which the Martian winds blew about the sand, but even though these Aeolian effects, as he called them, were of interest in explaining the splotchy, kaleidoscopic appearance of the planet from a distance, they were, after all, a sort of halfway house between geology and meteorology and as such were not his primary interest. Besides, at present there wasn't much wind to speak of. Sagan was in a dilemma common among theoreticians, who often find themselves becalmed when the data begin to arrive. They have to wait for enough data to accumulate so that they can generate more theories—or adjust the old ones to the new facts. While Sagan's aggressive persistence had been what chiefly annoyed his colleagues, there were those who missed his combativeness now. Much as they might strive for objectivity, the scientists were coming to recognize that they were no different from other human beings—that, indeed, they, like other people, tended to have strong opinions, and that these opinions might be part of the scientific process.

In this period Sagan was hardly idle. He was all attention, continually going over the evidence in his mind as it came in, asking questions, and occasionally even nudging a colleague along the path he thought right. And although he admitted to a certain initial disappointment at the lack of macrobes, he had not thrown in the towel. As he often did when the going got sticky, he resorted to a negative argument. He had been pointing out for some time, he said, that the landing site had been chosen, for reasons of safety, for its blandness; even though there were no visible signs of life at this spot—the cameras took in only one twenty-millionth of the surface of Mars—there were plenty of other places, and more interesting places at that, where life—*visible* life—might lurk. But all the scientists, Sagan included, were about ready to turn their attention from macrobes to microbes. On Earth, even, there were spots where life was not immediately visible. "For example," Sagan said, "in the great Peruvian desert, as far as the eye can see, there is nothing alive whatever; and yet the Peruvian desert is loaded with microorganisms." Whatever the case with the great Peruvian desert, on the one at Chryse the hunting ground would shift from the acreage in view of the cameras to the scoopful of dirt that would shortly be dumped into the biology hopper.

On sol 8—the eighth day after the landing, corresponding more or less to July 28 at J.P.L.—the sampler arm extended and then dropped to the ground. The trenches for the first sample, according to the preprogrammed sequence, were supposed to be dug directly in front of the lander, but that area turned out to be especially littered with rocks; accordingly, the program was changed and the arm stretched out to the lander's left, toward a bare patch in front of a rock the geologists had named Shadow. According to one of the geologists on the special properties team, which worked with the sampler arm, the rocks around the spacecraft were so pervasive that if the scientists and engineers in Pasadena had been unable to make decisions more or less on the spot, and had had to rely on the preprogrammed sequence as the Russians, who lacked the flexibility, did, there was only a 15 to 20 percent chance that the sampler arm would have been able to gather any dirt at all. As it was, a number of geologists and engineers feared that the sampler arm would hit Shadow Rock, just beyond where it was supposed to dig.

The sampler arm pushed into the ground, digging a small trench, and dumped the soil into the hopper that led down to the biology package below decks. The red soil dribbled through a coarse sieve at the front of the scoop and then through a finer one inside the funnel at the top of the hopper. Then the scoop went back for another load of soil for the second hopper, which led down to the gas chromatograph mass spectrometer that would examine the sample for organic chemicals. The arm apparently failed to make the delivery, however—a mishap that would cause lengthy delays for Viking's organic chemists, a group not known for its composure. The next day, the scoop dumped soil down the third hopper, which led to the X-ray fluorescence spectrometer that would analyze the sample for inorganic chemicals.

The hopper for the biology package—the first of the three instruments to get soil, as it had the highest priority—was able to move its lower end in a circle around a small track, dumping soil into three small openings leading to containers of different capacities; when each container was full, a metal plate closed over the top and another at the bottom slid open, so that precisely measured quantities of soil trickled into each instrument's incubating chamber. The incubations started over the next 3 days, and almost as soon as they did, things began to happen.

That anything could happen at all was unexpected on at least two scores. First, because of the problems with the tests, no one was certain the instruments would work at all; indeed, two of the three instruments now on Mars were known to have anomalies, though the scientists in charge believed they could compensate for the defects when they analyzed the data. And second, of course, was the low chance that any form of life existed on Mars. Nevertheless, 3 days later, on the eleventh sol, Harold Klein, the leader of the biology team, was standing before a press conference at J.P.L. announcing what he called "important, unique, and exciting things."

When the soil entered Oyama's gas-exchange instrument, which would examine it in a rather broad-gauged way for signs of heterotrophic, or animal-like, activity, it dropped into an incubation chamber—a cylinder of stainless steel about 1 inch in diameter and 2 inches high. The entire instrument was a conglomeration of tiny pipes, valves, and tanks stuffed into about a third of the cubic foot of the biology package. Toward its bottom was a revolving carousel that carried, in addition to the incubation chamber, a second chamber, called a dump cell, for surplus soil. (Any soil in the hopper eventually had to be taken into the instruments as there was no other way to get rid of it, and there would be many times when a scientist didn't want it.) The carousel moved the incubation chamber with the soil inside until it was underneath a gasket; a spring jammed it up against the gasket, effectively sealing it. Nutrients and gases would enter the chamber through small pipes inside the gasket. Later, as the incubation progressed, gases—different gases, it was hoped—would be tapped from the chamber through other pipes for analysis. The nutrient contained water and the 19 amino acids that are most commonly found in meteorites and are therefore expected to be prevalent elsewhere. Oyama hoped a Martian organism would find something it liked in this rich mixture, which some of the other biologists called "chicken soup" and which Norman Horowitz had dubbed Oyama's Pharmacy.

Oyama, a cheerful, frequently smiling man, placed the chances for life on Mars far higher than anyone else on the team—at 50 percent. (Klein's 2-percent chance was more typical of the rest.) Oyama had been extremely optimistic about the existence of life on other planets ever since he had begun to work for NASA in 1959; he had searched for life in, among other places, the rocks the astronauts brought back from the Moon; and in 1978 another instrument he had designed would fly to Venus to look there for traces of water vapor and other compounds favorable to life. Among the Viking biologists, now that Vishniac was gone, Oyama's views were considered extraordinarily upbeat. Horowitz, the team's pessimist with respect to life on Mars—he is a trim, white-haired man with a square-cut face a little like a fox terrier's—seemed to oppose Oyama almost as a matter of principle; he was forever bringing up reasons to be dismayed by the chicken soup. Because of Viking's electronics, the temperature of the lander, and thus of the entire biology package, was kept at about 10° C, and therefore Horowitz and the others could not carry out their incubations at normal Mars temperatures, which even in summer at Chryse Planitia averaged about −50° C. A way around this might have been found—but because of the water in the chicken soup, as well as in Levin's nutrient, the biology package, to Horowitz's distress, could not be kept at below-freezing temperatures. (As long as that was the case, Horowitz had decided to send along a small amount of water himself, but it was not essential to his experiment, and he wouldn't have minded sacrificing it for a subzero incubation.) The presence of liquid water meant the pressure had to be kept at non-Martian levels, too, for otherwise the water would boil away. Any water on the surface of Mars, above molecular quantities too meager to freeze or evaporate, would be ice, not liquid, Horowitz pointed out. Horowitz had frequently insisted that any organism coming in contact with Oyama's chicken soup would instantly either drown or burst. He felt this would be true even of the vapor that Oyama would use in his first experiment. Oyama, though he tried to remain imperturbable, clearly resented the criticism; he thought he had arranged things so that the organisms wouldn't drown or burst.

In Oyama's experiment, the soil lay, not on the bottom of the incubation chamber (as it did in the two others), but in a stainless-steel tray whose bottom was porous. Oyama injected a very small amount of his nutrient from below, so that it wet the bottom of the chamber but did not touch the soil. However, water vapor gradually spread through the soil to create gradations of moisture. Oyama felt that if the Martian organisms liked the water they might flourish and reproduce. It would take time, though, before there were enough of them to generate the gases his instrument could detect—carbon dioxide, carbon monoxide, methane, hydrogen, nitrogen, oxygen, nitrous oxide, and hydrogen sulfide—in sufficient quantities for his sensor, a gas chromatograph, to detect. At the very least, Oyama thought, it would take a few days, and he was prepared to wait weeks or even months. Yet his first reading, received 2.5 hours after the nutrient had been injected into his incubation chamber, showed that 20 times as much oxygen had been released as could be expected from the Martian atmosphere in the chamber. There were no other anomalies to speak of, except a slight rise in carbon dioxide over the amount that could be accounted for by the atmosphere. A reading made a day later showed only about a 30-percent increase in oxygen. Clearly, most of it had been produced before the first reading, and thereafter the rate of release faded as quickly as it had come.

The soil in Levin's label-release instrument, which would also be examined for signs of heterotrophic, animal-like activity, though by a narrower, more specific method, was incubating, too. On its carousel were four incubation chambers, each smaller than Oyama's single incubation chamber. Levin's instrument, like Oyama's, looked for signs of heterotrophic metabolism by adding a nutrient to a sample, allowing it to incubate, and then seeing if there was any change in the gases in the chamber. In Levin's instrument, though, the nutrient was injected from above, directly onto the soil in the bottom of the chamber. Levin's nutrient, unlike Oyama's rich chicken soup, was a relatively weak mixture of only seven synthetic amino acids and carbohydrates, which he called, collectively, his *substrates*—the technical name for substances acted upon by the enzymes of living organisms. The acids and carbohydrates had been synthesized artificially because they had to be made with carbon 14, a radioactive isotope of carbon, which could be easily identified later. (Scientists call this use of carbon 14 *labeling*.) The idea was that if an animal-like organism ate one of the substrates and metabolized it, breaking it down to create new protoplasm and energy, the organism would exhale some of the carbon in the form of one or another of the carbon gases, the most likely one being carbon dioxide. Then the gases in the chamber's atmosphere would pass through a Geiger counter. If any of the gas turned out to be radioactive, that would indicate that the radioactively labeled carbon from one of the substrates had been metabolized and had then passed into the atmosphere—a situation presumptive of life. Since Levin's instrument detected only radioactive carbon gas, as opposed to the half-dozen or more gases that Oyama's looked for, it was considerably more specific—an arrangement that had both good points and bad points. Within its own specialty, it was many times more sensitive than Oyama's system, for a Geiger counter could detect quantities of radioactive carbon 1,000 times smaller than the amounts of gas Oyama's gas chromatograph could detect. Consequently, Levin's instrument needed less soil than Oyama's and—in theory, at least—should get results a lot sooner.

The first 9 hours of Levin's data came down all at once and were delivered in the form of computer printouts at 7:30 P.M. on July 30, 2 days after the soil had been dumped into one of his instrument's incubation chambers and less than 1 day after the nutrient had been injected. Almost immediately, the Geiger counter had begun detecting radioactive carbon in the chamber's atmosphere; the count rose quickly, from a starting point of about 500 counts a minute (the zero level, in effect, because radioactivity from the lander's radioisotopic generators made a constant background noise) to 4,500 counts a minute over the background level, which it reached at the end of 9 hours. This was faster than most biologically rich soils in Earth would react. "The odds were overwhelming that nothing would happen at all," Levin said, after he had plotted the data points on a graph. "And when we saw that curve go up we all flipped. We knew we had a sample, we knew we had a nutrient injection, and we knew something was happening there." Levin and a number of his associates sat down and solemnly signed the first page of the data printouts, very conscious of the fact that they might be commemorating the first indications of life on another planet.

With two of the three biology experiments on Mars giving active responses, some of the members of the biology team held a press conference at the Jet Propulsion Laboratory on sol 11

(July 31), 3 days after the scoop dropped the soil into the hopper. (The third instrument in the biology package, Horowitz's pyrolytic release, had not yet returned its results; nor was there any information on organic compounds yet from the GCMS because of the mishap on sol 8.) The biologists walked from their offices across a broad, sunny plaza to the von Karman Auditorium, a one-story building next to J.P.L.'s front gate. J.P.L. looks more like a Spanish hill town than like a NASA base, for its 100-odd buildings, together with many tall green pine and eucalyptus trees, cascade down one of the foothills of the San Gabriel Mountains; different levels are connected by steep flights of steps. (Strictly speaking, J.P.L. is not a NASA base but part of the California Institute of Technology, which runs it for the space agency.) Just off the auditorium lobby was a working model of the lander sitting in a box of sand contoured to duplicate the ground immediately in front of the lander at Chryse; even the rocks were duplicated in Styrofoam; engineers had been working overtime recently on this simulation as they tried to figure out what had gone wrong with the soil delivery to the GCMS.

In the auditorium, members of the biology team were seated on the platform under bright television lights. They were flanked by another replica of the lander, along with a replica of the Viking orbiter, its instruments protected (as they were in space) by a blanket of aluminized Mylar. On the platform, Klein, the team leader, said that Oyama's gas-exchange experiment had provided "at least preliminary evidence for a very active surface material" and that the response from Levin's label-release instrument looked "very much like a biological signal"; however, he added, the results "must be viewed very, very carefully."

Klein and the other biologists had every reason to be cautious. It would be nice to find life on Mars—the discovery would be easily as important as any in history—but no one wanted to make a fool of himself. All were aware of the predicament of the scientist who had thought he had found signs of life on Mars during the Mariner 6 and 7 missions, which had flown by the planet in 1969. What had happened was that he thought the craft had discovered methane and ammonia—two biological products—in the atmosphere of the planet and had announced during a press conference at J.P.L. that there very likely was life on Mars. "He'd misinterpreted his data," Horowitz, who had been present then, told a reporter. "None of us want to get caught like that. We might not be as graceful getting out of it." That scientist had simply retracted his data. There were other such stories, too. "I ascribe them to a common syndrome of Mars observers to deceive themselves," said Horowitz, who of all the Viking scientists was perhaps the most interested in what motivated his fellow scientists. He was a close friend and colleague of Bruce Murray's both at Caltech and at J.P.L., which Murray now directed; and the two scientists had developed their skepticism about life on Mars, and their ideas about wishfulness, together. "It may still be true of some people. The history of Martian exploration is an interesting exercise in the psychodynamics of science. Or scientists." In Horowitz's view, there was a distinct possibility that history would repeat itself, over and over again. He felt that the nature of Oyama's nutrient was only the latest example of the wishful thinking underlying the search for life on Mars— however objective the scientists might all try to be. It was a frustrating time, and the biologists—Horowitz among them— would bend in all sorts of different directions.

After Klein had shown some slides, including one of Oyama's spectacular oxygen peak, the first question asked by a reporter was whether the oxygen was being released by photosynthesis—in other words, by some sort of plantlike organism in the course of forming carbohydrates (or something similar to them) in a process involving carbon dioxide, sunlight, and chlorophyll (or some Martian substitute). He asked the question despite the fact that Oyama's instrument was designed to look for animal-like life. (On Mars, anything seemed possible.) Klein was quick to dismiss the oxygen peak as a sign of plantlike life because Oyama's instrument had no light to bring about photosynthesis. In a conversation afterward, Klein was also quick to dismiss the oxygen peak as a sign of animal-like life. Animals don't release oxygen. Besides, there hadn't been enough time for any oxygen-releasing Martian bugs—assuming that in this regard they resembled terrestrial ones—to reproduce and generate detectable gas changes, he said, and he pointed out that because Oyama's instrument could not detect amounts of gas as small as either of the others could, no one had expected it to detect a response as soon as Levin's did. An immediate response, however, was most likely to be chemical, Klein explained, because if the right chemicals are present a reaction is apt to occur immediately. "We believe there is something in the surface, some chemical or physical entity, which is affording the surface material a great deal of activity and may in fact mimic—let me emphasize that: *mimic*—in some respects, biological activity," Klein—who, of all the biologists, was perhaps the best at explaining things to the press—had said at the outset of the press conference. It appeared that what a number of scientists had feared would happen—that the biologists would have difficulty interpreting their data—was indeed happening. At first glance, it looked as though the Viking geologists had been right when they had suggested that the biology instruments were too new and untried, that too little was known about conditions on the planet for the experiments to return anything other than ambiguous evidence, and that the resultant uncertainties and disputes over whether there was life on Mars would be what one of them called a "horror show."

The possibility that the release of oxygen in Oyama's instrument—which had been designed with the expectation that gas releases would be evidence of life—had a chemical cause further confused the search for life on Mars, a search that had already suffered more than its share of such confusion. The chemical reaction that Klein, Oyama, and many others had in mind as an explanation for the oxygen peak—it had been proposed an hour before the press conference at a meeting of the biology team by two members of the molecular analysis team who had been invited to attend, John Oro and Klaus Biemann, and quickly approved by Oyama—was the cleaving, or loss, of oxygen atoms from some basic type of unstable oxidant. Of these there are three: peroxides are compounds that contain somewhat more oxygen than they can handle and, under certain circumstances, give up their excess of that element easily; superoxides contain even more oxygen than they can handle and give it up even more readily; least stable are ozonides. A fourth form of oxidant, the oxides—such as the iron oxides, which are probably what turn the Martian rocks and dust a rusty red—contain just enough oxygen to be chemically very stable, and for that reason they were not considered likely candidates for the cause of Oyama's oxygen peak.

Because the reaction in Oyama's instrument had been particularly brief and intense, Klein, Oyama, and others were looking hard at the superoxides, such as those made with calcium or zinc. (They also briefly considered the ozonides.) When water is added to either a peroxide or a superoxide, it loses much of its oxygen very rapidly, and this was quite probably the cause of what was coming to be called Oyama's oxygen peak. The water that caused the trouble, they believed, was in the chicken soup, which had just been injected. Even though all that reached the soil was water vapor, the surface soil of Mars was so dry that any superoxides would have released their oxygen with an almost explosive force. But the quantities involved were too small to damage Oyama's instrument; the oxidants represented only a few parts per billion of the soil.

As soon as the scientists thought about it, they realized that the presence of oxidants on Mars was theoretically likely, though they couldn't prove the theory. While it is true that water destroys most oxidants, tiny amounts of water are essential for oxidant formation, and tiny amounts of water are what Mars has. If the entire amount of vapor in the atmosphere above the lander were squeezed out of the air—something that would never happen—it would form a film 0.00156 inch thick. (The scarcity of water on Mars did not rule out life, which even on Earth can get along with practically none, at least in a dormant state; and conceivably life on Mars was dormant during dry spells, as Sagan had once suggested, or there were life forms that required almost no water even when active. Besides, there could be additional water frozen underground.) In the Martian atmosphere, the intense ultraviolet radiation from the Sun would cause a reaction in which the water H_2O lost one of its hydrogen atoms, leaving a paired oxygen and hydrogen combination (OH), which chemists call a free radical—meaning that it is a fundamental constituent in the formation of other molecules. Alone, this constituent could well be in the soil, contributing to Oyama's oxygen peak. The radical is highly reactive, though, and would most likely be ancestral to whatever oxidants might exist on Mars. For example, while still free in the atmosphere it could easily react with another, identical OH radical to form the molecule H_2O_2—or hydrogen peroxide, the simplest of all the peroxides. Many scientists considered its presence on Mars very likely. Hydrogen peroxide is a clear substance often used on Earth as a bleach or an antiseptic. On Mars, it would presumably be mixed in with the red iron oxides (hematite, probably though there might be others as well) that apparently coated the rocks and dust—as were any other peroxides and superoxides. Alternately, one or more OH radicals, having very likely settled on dust particles while they, too, were floating about in the atmosphere, would attack any metals present, oxidizing them; later, the oxidant-rich dust would settle on the ground. At this stage, however, it was not known to what extent these things had happened, despite the fact that Oyama was looking hard at superoxides to explain his oxygen release; it was not known whether there were just oxides on Mars (as seemed certain from the reddish color of the dust), whether there were peroxides, too, or whether there were superoxides as well.

Even in the infinitesimal quantities supposedly present on Mars, oxidants are lethal to organic compounds and to most of the primitive forms of life. Oyama, having put the odds for life on Mars as high as 50:50, now dropped them to 1 or 2 percent. ("Most people would say there were no leprechauns in Ireland, and be done with the question," said one exasperated Viking geologist, who clearly thought the likelihood that Mars was covered by a poison put the question of life there in the leprechaun area. "But a biologist could not make that categorical statement. Rather, he would say, 'The likelihood that there are leprechauns in Ireland is one billionth of one percent,' and then he would spend the rest of his life trying to change the number of decimal places.") Oxidants are lethal in part because an organism will always add a little water to them, causing a forceful release of oxygen; but more important, the newly released oxygen is itself lethal to most organic compounds. Free oxygen is a highly profligate element, which forcibly joins itself to certain other elements, even if that means breaking up existing compounds; in particular, it destroys the basic bond between carbon and hydrogen, linking up with the carbon to form carbon dioxide (CO_2). Oyama had detected in his instrument, in addition to the oxygen release, a slight excess of carbon dioxide above what was expected from the atmosphere, and a number of scientists believed that reflected, among other things, that the oxygen was attacking the carbon in his organic nutrient.

The statement that oxygen is lethal to most organic compounds may sound strange because animals breathe oxygen and would die without it; on Earth, plants and animals thrive in an atmosphere that is 20 percent oxygen. This adaptation is the result of both planetary and biological evolution, which scientists increasingly think go hand in hand. One of the oldest known forms of life, which developed over 3.5 billion years ago, are methanogens, part of a newly identified biological kingdom, archebacteria, which are neither plant nor animal but are thought to be collateral relatives of both, and which inhabit such anaerobic spots as the mud beneath deep hot springs in Yellowstone National Park; they die when they are exposed to oxygen. When the Earth and Mars were young, both had atmospheres very different from their present ones—with more hydrogen than they have today. A planet must have hydrogen for life to develop at all. Hydrogen joins with carbon (a reaction chemists call *reduction*, meaning the combination of hydrogen with other elements) to form in the basic organic bond.

Hydrogen was the most abundant element in the solar nebula from which the solar system condensed, and the gas was carried to the planets during the accretion process that built them up. Some of it was in the Earth's atmosphere to begin with, though in time it escaped to space, for hydrogen is the lightest element. Part of the Earth's original hydrogen, deep inside the planet, was outgassed later—much of it already having joined, in the Earth's interior, with oxygen to form water that rose as vapor into the atmosphere, where the ultraviolet light of the Sun broke it down; some of that hydrogen escaped into space, but the oxygen, being heavier, sank back into the atmosphere, in the form of the OH radical, of oxides, or of free oxygen. Gradually, the atmosphere of both the Earth and Mars became more oxidative.

Early in the Earth's history, when the atmosphere was oxygen-poor, the only forms of life that flourished were anaerobic bacteria such as the archebacteria. The loss of hydrogen and the transition of the Earth's atmosphere from oxygen-poor to oxygen-rich were undoubtedly the most cataclysmic challenges life on Earth has ever faced, giving rise to the basic forms of life that we know today. Very likely, most early forms of life here were wiped out. A few, like the methanogens, which lived in airless places underground, weren't affected. And a few others—those which had advanced sufficiently to be capable of making the change—adapted slowly to the new environment; these were the earliest unicellular plants and animals. They developed an enzyme called catalase, rich in iron, that rapidly broke the oxidants down. (Catalase, Sagan pointed out in a conversation at this time, is found in all animals and plants—possibly harking back to a time when oxidants were the major threat to life.) They also developed ways to use the destructive powers of oxygen for their own metabolic purposes: animals learned to derive energy by oxidizing the carbon in their bodies to carbon dioxide, and plants learned to derive energy, reductively, by photosynthesis. With these adaptations, plants and animals evidently transformed our planet, the plants adding abundantly to its oxygen supply.

On Earth, most of the oxygen is in the atmosphere, while on Mars it was evidently on the ground—a possibility no one, despite the rusty color of the ground, suggesting iron oxides, had paid much attention to before Oyama's data came in, and one that many Viking scientists found fascinating, for it suggested a point at which the history of the two planets diverged and stimulated some thought about the evolution of life.

It is very possible that on Mars life never started at all. Or perhaps it did get a start, but was later wiped out by the increasingly oxidative environment; in that event, Viking could conceivably still detect traces of it. Or perhaps some forms persisted underground without changing, like the methanogens; if plants and animals had never evolved on Earth, there could still be methanogens here, on a planet that might otherwise be very much like Mars. Or perhaps some forms of life made the additional step of adapting to the new environment—not in sufficient numbers to have altered the planet, but perhaps in numbers that still might be detected by Viking. Oyama's data, though they hadn't ruled out life, had clearly redefined the search for it on Mars.

Levin's curve, which had been going almost straight up for the better part of a day, began turning more and more to the right as the speed of the reaction diminished. By the end of the second day, it had reached a level of about 8,500 counts a minute; it would continue a slow climb, to about 10,000, before leveling off a day or so later. Levin, still on the crest of the wave, or the curve, did not seem to give much weight to the possibility that his results might be attributable to something other than biology. And why should he? He had designed his instrument so that an active signal, such as he was receiving, would be indicative of life. He had every reason to believe his instrument was working well. The curve might have been produced by a stowaway microbe from Earth, though this seemed almost impossible because the entire Viking lander had been heat-sterilized inside the bioshield, in which it had subsequently been launched, unopened. Levin, who had previously been less outspoken on the chances for life on Mars than Oyama had started out being—let alone Sagan—became increasingly willing to believe that life might exist there.

Levin, a tall, tanned, impassive man, had been in the life-detection business longer than any other member of the biology team. It was as far back as the early 1950s that he had conceived the idea of using radioactively labeled carbon concealed in a nutrient and exhaled by microorganisms to determine whether the number of organisms in municipal water supplies had reached the danger point. Since taking his invention to NASA in 1960 (no municipality had bought it), Levin, with one version of his machine or another, had detected life in some of the bleakest spots on Earth— Death Valley, mountaintops—and partly for this reason the possibility that he had discovered life on Mars did not entirely surprise him.

Klein's response to Levin's results was different. "I know it's impossible not to ask, or not to focus on the question, whether this is or is not a biological response," he said at the press conference. "At present, there's no way you can rule out the data as being due to biology. However, let me say, if it *is* a biological response, *if* it is, then it's a stronger response than we have seen with fairly rich terrestrial soils, and it would also imply that microbial life on Mars is more highly developed— more intense—than it is on Earth." The terrestrial soils in question, from an area southwest of Lake Tahoe, in the foothills of the Sierras, had taken 168 hours to reach the count the Martian soil had reached in only 40 hours.

Klein and others pointed out that a biological curve should start out slowly, as the organisms got used to the nutrient, and then should rise with increasing rapidity as the organisms multiplied—a geometric growth rate, reflected in what scientists call an exponential curve. Levin, for his part, pointed out that his curve was not a typical chemical one either. If the reaction was chemical, he said the curve should start off much more quickly and continue upward at a steady rate, then level off quite sharply and drop when the nutrient was used up.

The scientists all agreed that Levin's curve was neither typically biological nor typically chemical. It was steeper than most chemical reactions. Klein and others were put in mind of the rapid release of oxygen they had seen in Oyama's results;

indeed, they were beginning to think that the same sort of thing was very likely happening in Levin's instrument—that the water in Levin's liquid nutrient had also caused oxidants in the soil to give up their oxygen. Levin's instrument could not detect oxygen directly, but oxygen would immediately attack the radioactive carbon in his organic nutrient, combining with it and causing the spurt of radioactive carbon dioxide reflected in Levin's curve—in fact, mimicking the very metabolic process his machine was designed to look for. Since Oyama's instrument had also detected a release of carbon dioxide (though not an excessive one), which suggested that the carbon compounds in his chicken soup had been oxidized, these more conservative scientists felt they had a strong case for oxidants.

Levin did not concur with this line of thought. If an oxidant was responsible, he said, then the oxygen, being such an intrusive, aggressive element, should have gone on combining with the radioactive carbon in his nutrient until the carbon was all used up; instead of stopping at around 10,000 a minute, the count should have gone much, much higher, until it approached 272,000 counts, which was the total amount of radioactive carbon in a single injection of his nutrient. The 10,000 count was under the 16,000 that would be expected if the reaction involved only one of the seven substrates—the seven ingredients in Levin's nutrient. The one that suspicion immediately fell upon was formate, a variety of formic acid; because it was the simplest and the most likely to exist on a primitive planet, Levin and others had thought it the most likely to be favored by a Martian organism. Furthermore, Levin pointed out, living things have a greater specificity and economy in their chemistry than nonliving things, and so organisms, far more than oxidants, were likely to pick one substrate over another. Sagan, coming to Levin's support, said contemptuously of Oyama's oxidant: "Your superoxide, or whatever it is, oxidizes the formate, but it can't oxidize anything else! That sounds funny, given a medium so rich in easily oxidized compounds! Could it instead be a finicky organism— one that is fussy about its food?"

The question of organisms versus oxidants was not going to be resolved easily. The dilemma sounded more and more like the "horror show" certain Viking geologists had predicted when they had suggested the biologists might not be able to "read" the data from their instruments. Indeed, at this time one geologist said: "The problem the biologists are having is exactly what we worried about—they don't understand the nature of the soil." In addition, the dilemma simply underscored the fact that the instruments, however complicated in their many mechanical respects, were very primitive when it came to attempting to detect life remotely—something that had never been done before. The very complexity of the question of what is alive, and what is not, made the instruments seem all the more clumsy. Sometime later, when Joshua Lederberg was asked why the scientists had not sent to Mars one single instrument that could tell whether the results they were seeing were chemical or biological, he replied: "The only instrument I know of capable of doing that would be something like a divining rod."

The ambiguous results were also a consequence of something that, as biologists have recognized increasingly over the last century, is a precondition of the biochemical theory of the evolution of life; namely, that biology is chemistry. At a panel discussion a few days after the press conference, Klein observed that he was reminded of the great 50-year debate—it lasted up until 100 years ago—about whether the transformation of grape juice into alcohol was a biological or a chemical phenomenon. "It took a long time for the scientists of the day to adduce the appropriate proof that it was a biological process— fermentation—and, of course, the biological part of the process was carried out by chemicals," he said. "Let's not separate biology from chemistry. The thing that's going on even in organisms is chemistry, and it's only a question of whether the chemistry we're looking at is inside a living system or whether it's outside, or exclusive of, a living system."

Speaking on the same panel, Joshua Lederberg took Klein's point a step further. He was reminded, he said, of work done 50 years ago by Otto Warburg, a German biochemist, who laid much of the groundwork of our present understanding about how oxidation is done within the cell. (There was something about the results from Mars that forced the more reflective biologists to look for analogies in the early study of biology, when the most basic research was being undertaken.) Warburg had used charcoal impregnated with iron salts to catalyze the oxidation of amino acids in the presence of oxygen, liberating carbon dioxide in a manner that Lederberg felt was not a bad model for what might be taking place on Mars. "And here is the complexity," Lederberg went on. "Warburg was using this process as a model of what's going on in the cell. Now, with Viking we're not able to decide, simply on the basis of that catalytic activity, whether we are looking at a primitive inorganic process or whether this process has been incorporated into cells and become part of a living system." Lederberg speculated that there might even be a stage in the early chemical evolution of a planet when it was able to generate the metabolism of organic substances. He did not push the idea any further, though Sagan later would.

If it had been difficult for scientists 100 years ago to conclude that fermentation was a biological rather than a chemical process when they had the grape juice right in front of them, the difficulty of making a comparable deduction was vastly increased when the sample was a couple of hundred million miles away, and the laboratory, an intricate one, had been programmed years before with only a very few alternative protocols. Furthermore, if there were organisms on Mars, they would probably be unlike anything biologists were used to detecting on

Earth. In the face of all these uncertainties, the members of the biology team fell back on a procedure they had agreed upon long ago—that before they concluded there was biological activity on Mars they would have to disprove all other possible explanations of the data. As Klein put it, they would have to be *"forced* into the biological explanation." Biology, after all, was not just another branch of chemistry but its most rarefied branch and its pinnacle.

Despite these good intentions, the arguments began right away. Oyama was convinced that Levin's results were caused by oxidants, too. The fact that his own instrument had detected more than the expected amount of carbon dioxide in addition to his oxygen peak was, as far as he was concerned, all anyone needed to know. Levin, however, needed to know more. If the results of the two instruments were caused by oxidants, why had the production of carbon dioxide gone on so much longer in his own instrument than in Oyama's? Oyama retorted that obviously this discrepancy was attributable to the fact that Levin's instrument, with its radioactive carbon labeling, could detect far smaller quantities of this gas. Horowitz, for once jumping to Oyama's defense, pointed out that Oyama's nutrient contained an agent that absorbed oxygen, ascorbic acid, but that Levin's did not. They all agreed that the argument was getting tenuous; other things, aside from oxidation of his nutrient, might be contributing far more to Oyama's carbon dioxide production, such as the release from the soil of carbon dioxide adhering to it. Levin attempted to turn the others' argument back on them by suggesting that a small fraction of Oyama's carbon dioxide might be caused by autotrophic, plantlike life—at least, the possibility shouldn't be ruled out. Certainly with respect to his own instrument, Levin insisted, the simplest way to account for the carbon dioxide production was biology. Perhaps the production of this gas in small quantities *was* continuing in Oyama's instrument, which was too insensitive to detect it.

Levin indeed had replaced Oyama as the member of the team who pushed hardest to keep the door open to the idea of life on Mars—a task he would perform diligently, though without the flourish of Sagan. (Not the least interesting aspect of Viking was that, as events moved away from Sagan and Murray, others rose up to fill their roles in the argument—almost as though science by its very nature demanded conflict, creating antagonists, or protagonists, as needed.) He was not convinced that Oyama's oxidants even existed. A little later, a scientist announced that he had managed to duplicate Levin's spectacular curve by pouring Levin's nutrient onto soil coated with an oxidant. Levin's response was indignant. "It isn't enough to stand up at a meeting and say, 'When I pour liquid nutrient on oxidants, I get oxygen coming off' or 'The oxygen then oxidizes the nutrient,' " he said. "Of course you will! You must go on to prove that there *are* oxidants on Mars and how much of them and whether they will behave the same way on Mars that they do in a laboratory on Earth!" There was not one jot of direct evidence for the oxidants—just a mishmash of indirect evidence, Levin said. Levin was a more aggressive debater than most of the others. Klein and Oyama, undeterred, were still thinking in terms of superoxides; if these were the explanation for Oyama's oxygen peak, as they thought, the simplest assumption was that they explained Levin's results as well. The behavior of oxidants was not a subject biologists normally paid much attention to—indeed, organic chemists, not biologists, had initially brought oxidants into the discussion. Now that a knowledge of oxidants was essential to confirming or eliminating the presence of life on Mars, the biologists were boning up on them as best they could.

On sol 16, Oyama injected more nutrients into his soil, and a day later Levin did the same, as if they were putting a fresh worm on a fishhook. (Indeed, at around this time Klein compared the lander to a boatload of fishermen, two of whom had received nibbles—though it was not known yet whether they had lured a fish or snagged a beer can.) Oyama squirted in a large amount of his chicken soup this time, and from above, so that the nutrients would now come into contact with the soil. The idea had originally been that if water vapor was not enough to perturb any Martian organisms, perhaps the nutrients would be, but now that Oyama was thinking in terms of oxidants he felt that the second injection might help him plot how much oxidant was present in the soil. There was no new release of oxygen. This result, he thought, favored chemistry, for if organisms had been responsible for the oxygen peak, the liquid injection should have caused another burst of activity. Possibly the oxidants had all been used up the first time. Sagan, who was not so quick to dismiss the possibility that Oyama's oxygen peak was biological in origin, took another view. What if the Martian organisms hated water even in the form of humidity? In that case, the oxygen peak Oyama had first detected might have been the death shrieks of the organisms gasping in the unaccustomed mist, and by the time of the next injection they would all have been dead, so there would be no more release of oxygen.

After Levin's second injection, there was no sign of increased activity, either; instead, his curve, which had remained almost level at a count of 10,000, rose briefly and then dropped to a count of about 8,000, where it remained. Everyone agreed that the initial jump had been caused by a pressure surge following the injection and hence was without significance. Those who favored chemistry argued that once again Levin's results were compatible with Oyama's, and since Oyama's results were quite likely caused by oxidants, this was further indication that Levin's were, too. Horowitz went so far as to explain how the oxidant theory could account for Levin's drop. In both instruments, he said, all the oxidants would have been used up after the first injection. Then, since carbon dioxide is soluble in water, the new humidity simply absorbed some of the radioactive carbon dioxide, causing the drop. Levin was incensed. The moisture would have absorbed some of the radioactive carbon dioxide anyway, regardless of whether it had been produced biologically or chemically, he said. If the activity was biological, Horowitz asked, why didn't the curve start back up after the drop? "That could mean that the bugs are dead—killed by the vapor of the first injection," Levin said, invoking Sagan's argument. Over the next 6 days, it seemed to Levin's associate, Patricia Ann Straat, that the radioactive count was slowly beginning to rise. The incubation ended before she could tell whether the curve was about to take off exponentially, as it would if organisms were present, or whether the rise would continue at all. She and Levin both hoped they would be able to repeat the experiment and let it go on much longer.

"If we just had Levin's experiment alone, we might very cautiously be announcing right now that there might be life on Mars," Klein said. "But with Oyama's experiment there too, the argument is pushed in the other direction. We don't know yet which way Horowitz's results will push us."

A couple of days before Levin's second injection, Horowitz received his first results. In any round of experiments, he was always the last to get his data, because his instrument did not make its measurements until the incubation was over—unlike the two other instruments, whose incubations were monitored throughout, in ever-lengthening graphs. Horowitz's results came in the form of two separate numbers, and when he got them, that was that. His instrument differed from the others because he was looking for reductive activity of the sort performed on Earth principally by plants, which use water and sunlight to take the carbon dioxide from the air and photosynthesize it, retaining the carbon as part of their structure. The instrument should not be sensitive to oxidants. Since the process Horowitz was studying was the reverse of what the others were investigating—the oxidative one that animals use—his instrument, despite the fact that it used radioactively labeled carbon, like Levin's, was in many other respects its mirror image.

For example, the two other instruments carried out their incubations in the dark, but Horowitz, because he was looking for plantlike microorganisms, needed sunlight, and he simulated this with a xenon lamp; it radiated most wavelengths of light, including both visible and invisible ultraviolet, which reach the Martian surface—though the shorter invisible wavelengths were blocked by a filter. Horowitz had three incubation chambers on his carousel. Once the soil was in a chamber and the chamber was sealed, the Martian atmosphere that happened to be inside—principally carbon dioxide, with traces of carbon monoxide and other gases—was augmented by a special preparation of carbon dioxide and carbon monoxide in which the carbon was in the form of radioactive carbon 14. In Levin's instrument, the radioactive carbon started out in the nutrient and wound up in the atmosphere; in Horowitz's it started in the atmosphere and would wind up inside microscopic plants—if any were there. The xenon lamp provided the energy to encourage photosynthesis. In theory, if there were plantlike microorganisms in the soil they would assimilate the radioactive carbon into their bodies.

Finding out whether this had happened required a two-step process that began after 5 days of incubation. Before the first of the measurements could start, however, the atmosphere inside the chamber, including the labeled gases, was flushed out. Some quantity of the radioactive gases might have been absorbed into the soil (this would depend upon the soil's alkalinity, among other things), so the first step was to heat the soil to 625° C—high enough not only to drive out the radioactive gases caught in the soil but also to destroy any microorganisms and to break down the organic material they contained. All this was swept into a golden U-shaped pipe known as the organic vapor trap—it was the column of a gas chromatograph—which separated the compounds by size. Any organic molecules larger than methane would be trapped inside the column, whose packing had an attraction for organic molecules, but most of the radioactive carbon monoxide and carbon dioxide that had been absorbed by the soil passed on through. (Since the column also had a small affinity for carbon dioxide, it had been saturated, shortly after landing, with the Martian atmosphere, which is mostly carbon dioxide, lest it trap any now.) These inorganic gases were then swept on into a radioactivity detector; this measurement, called the first peak, had nothing to do with life, but provided information about the atmosphere and the absorbency of the soil. After that, the golden pipe itself was heated to 700°—enough to break down any organic compounds trapped there. Some of these might have metabolically incorporated some of the radioactively labeled carbon during incubation. The released carbon would then combine with copper oxide present in the column to form carbon dioxide. The result of the measurement of this gas, called the second peak, would be indicative of possible biological activity.

Horowitz was extremely proud of his instrument, and he made no attempt to conceal the fact that he thought it superior to the two others and would just as soon have had them left behind. He got away with a certain bluntness, partly because he was one of the team's most distinguished members by virtue of his discovery that one gene governs the production of one enzyme. When, in the early 1960s, he had been working with Levin on Gulliver, the prototype of the label-release instrument, he too had been quite hopeful about life on Mars—this was before Mariner 4 had shown that liquid water could not exist on the surface of the planet. As he saw it, the underlying problem with instruments like Levin's and Oyama's, which were looking for animal-like life on Mars, was that since little heterotrophs could not prosper without nutrients and moisture there was always the risk that they would be offered the wrong nutrients or the wrong quantities of liquid. The beauty of looking for autotrophic life was that plantlike microorganisms needed only carbon dioxide, of which there was plenty on Mars, water (which existed there, too), and sunlight. Or a Martian organism might simply assimilate the prevalent carbon dioxide in some other way. Whatever they needed in the way of nutrient or water, the microorganisms would bring with them, in the soil. His own instrument supplied no nutrients, liquid or otherwise—though it had the ability to add minute amounts of water vapor. "My experiment is the only one designed to be purely Martian," Horowitz said shortly before his data came back. "The others can say their results are ambiguous—but not mine. We can be sure that if there's life on Mars, we didn't kill it. And because we don't need to add water we're relatively unaffected by the oxidant chemistry, too. Whatever the results, there's no way out."

Like Lewis Carroll's chess game in *Through the Looking-glass*, the search for life on Mars was fraught with certainties and rules that seemed to change from one moment to the next—a situation Horowitz fell victim to more precipitously than anyone else. His first-peak result—the one that had nothing to do with the presence or absence of life—was a count of several thousand per minute, and some such high count had been predicted, since it was expected to reflect the radioactive carbon gases absorbed by the soil and then released by the heating. Indeed, neither the press nor most of the biology team paid much attention to Horowitz's first peaks except to note that the machine was evidently working properly. But the second-peak result, which came a few days after the first peak and was to indicate whether or not metabolic activity was going on, got a great deal of attention, for its count was 96—a sign of significant activity. (Activity was officially defined as any count above 15. Actually, the minimum active count was not 15 at all, but one-fifth that, 3; Horowitz and his associates had raised the official active count, thus increasing fivefold the margin against the sort of error they feared most—and indeed, in a number of tests with sterilized soil, Horowitz and his associates had found a slight statistical spread in what should have been zero results.) "I was incredulous," Horowitz said, when asked whether he thought it strange that, with his relatively pessimistic view of life on Mars, he was the one whose instrument gave what appeared to be the most unambiguous indication of life. "You could have knocked me over with one of those Martian cobbles." He couldn't claim, as Oyama and Klein did of Levin's results, that oxidants were mimicking life, because he had used no water. Like Oyama, he had been forced to alter his position—he was in Oyama's situation, only in reverse, for Oyama, who had been the most optimistic about life on Mars, was the one who had found the strongest evidence against it. Many more reversals followed in the search for life on the planet known since the days of Lowell for its ambiguity.

Though Horowitz's count was more than 6 times as great as the official negative (and 32 times as great as the unofficial one), 96 was not in itself a large count, considering that a reasonably lively terrestrial soil had given Horowitz a second peak in the thousands, and even a sample of soil from the most inhospitable sort of place on Earth—a dry Antarctic valley—produced a reading of 106. Because the surface of Mars is drier and colder than the driest and coldest Antarctic valley, however, the very lowness of the count seemed to support the idea that it was generated by Martian biology. The 96 seemed to represent a much lower number of organisms than Levin's curve, which had aroused suspicion because it was so very high. Horowitz later figured that his count could have been generated by 1,000 bacterial cells—not many for the amount of soil in the chamber, but still a respectable number for what might be the first signs of life ever detected on another planet.

After the data arrived, Horowitz and a couple of associates spent several hours going over them to see, as Horowitz put it later, "whether there was any way we could talk ourselves out of it." They couldn't. The next morning, August 7, Horowitz presented his findings to the entire biology team; many of his colleagues were struck by his composure. He explained afterward: "If you're a scientist, you have to be willing to murder your own favorite ideas in the face of the data. If you are passionate about your beliefs, you can run into trouble." He was not sure that all his colleagues were so open-minded. For a man who had once said he thought the chance of finding life on Mars was close to zero, Horowitz was in something of a fix; he didn't want to lean too far in the other direction, either. Accordingly, when he addressed a press conference that afternoon, he declared: "I want to emphasize that we have not discovered life on Mars—*not*. However, the data point we have is conceivably of biological origin, but the biological explanation is only one of a number of alternative explanations that have to be excluded. We hope by the end of this mission to have excluded all but one of the explanations, whichever that may be."

Though Horowitz was managing to lean over forward and backward at the same time, he left little doubt about which explanation *he* thought would be among those ruled out; not even Horowitz had completely murdered his ideas. All the scientists were coping as best they could with the wishfulness that Horowitz especially knew was inherent in the search for life on Mars; some embraced the idea of life, others resisted it, and all clung as hard as they could to the scientific method or their differing view of it. Having to talk to the press clearly added to the strain, for Horowitz told the reporters: "I want to emphasize that if this were normal science we wouldn't even be here—we'd be working in our laboratories for three months more. You wouldn't even know what was going on, and at the end of that time we would come out and tell you the answer. Having to work in a fishbowl like this is an experience that none of us is used to, and you have to understand that you're in on the sorts of things that happen all the time in science—when hopeful data are obtained, one's hopes are raised for a while, but then, on further investigation, they turn out to be something else. That's the stage we're in. You're looking over the shoulders of a group of people who are trying to work in a normal way in an abnormal environment." Having reporters around caused all sorts of unusual strains. Long before the Viking scientists had submitted their official results for publication in *Science*, the chief scientific journal of an interdisciplinary nature in this country, that periodical had already received an unsolicited article, from someone who had no connection with Viking, which amounted to an accurate, scholarly treatise about the Viking discoveries; the author cited as his primary source articles in *The New York Times*.

Despite Horowitz's attempts to keep the lid on notions that he had discovered life on Mars, there was a definite undercurrent of excitement. Sagan, who had been in Washington proofreading the imaging team's official article for *Science* (its footnotes did not include the *Times*), came racing back to J.P.L. asking why all the ferment had happened while he had been away. Sagan's evident enthusiasm added to Horowitz's discomfort. He and the other biologists were tiring of data they couldn't explain.

Since the biologists would have to rule out all other explanations before being "forced," as Klein had put it, to the biological one, at the next day's session of the biology team (it met every morning at nine), they compiled a list of possible explanations for Horowitz's data. By the end of the session, they had a dozen alternatives—both chemical ones and ones they called "instrument anomalies." The latter were the easiest to check. One possibility was that Horowitz's radioactivity detector had malfunctioned and was counting at too high a rate. The following day, the gases remaining in the detector were to be blown out; if the detector was working properly, the count should immediately drop to near zero. It did, eliminating that worry. Another possibility was that there might have been some obstruction in the organic vapor trap, causing the radioactive gases from the first peak to remain there and create the high second peak. Shortly afterward, when the trap was blown out, the scientists searched the data to see if the detector counted a short burst of radioactivity. It didn't, and that meant the trap was in good order. In similar fashion, Horowitz checked almost all the possible malfunctions and found none. The biology package, in fact, was working almost perfectly—far better than it had during the integrated tests on Earth before the launch. The scientists thought this was because the engineers had developed greater proficiency in handling their instruments, but the engineers—who may not have wished to be reminded that there was once a time when things had gone wrong—asserted that, as the instruments had been designed to operate on Mars, they would, of course, work better there.

There was one possible instrument malfunction Horowitz could not check right away—a broken filter on his xenon lamp. In tests he and an associate, Jerry Hubbard, had conducted 4 years earlier, before there was a filter, it was found that the instrument sometimes detected high second peaks—apparent evidence of life—even when the soil being used was known to be sterile. This was a startling flaw in an instrument whose principal developer took pride in its lack of ambiguity. The flaw turned out to be of considerable scientific interest: Horowitz and Hubbard discovered that inside the instrument, under the ultraviolet light, radioactive carbon monoxide and slight amounts of water vapor in the atmosphere generated several small organic compounds on various soil surfaces. For example, the ultraviolet light caused photolysis of the carbon monoxide (CO) and the water (H_2O), and they generated formaldehyde $(HCHO)$. Other compounds generated in this fashion were acetaldehyde and glycolic acid. If such compounds were now being formed in Horowitz's chamber—employing the radioactive carbon monoxide in the chamber's atmosphere—they would be pyrolyzed and cause a high second peak. Because this synthesis of organic compounds could be happening outside his instrument as well as inside, Horowitz thought that Mars might well be coated with, say, formaldehyde. This was a totally new theory for the creation of organic compounds on a planet.

The fact that a machine for detecting life on Mars was capable of generating its own evidence might have been one of the most astonishing examples of wish fulfillment in scientific history if it had not been that Horowitz and Hubbard had discovered the danger, and had devised a means of averting it. Further study had shown that organic compounds were synthesized only under the shorter, invisible wavelengths of ultraviolet light—those below 3,200 angstroms, or almost at the borderline of visible light. When these wavelengths were eliminated with a filter, the instrument gave a negative response in tests using sterile soil. The filter was made a permanent part of the instrument. Now, though, the filter in the instrument on Mars might be broken, allowing the shorter wavelengths through, and that could account for the 96. The engineers did not think this likely; however, Horowitz could not be sure about it until he had done his next experiment, called the control, in which fresh soil would be baked and sterilized before incubation. If the result was a count higher than 15, the activity would be nonbiological and a broken filter would be a good bet.

There was an even more fundamental worry. What if on Mars the *longer* wavelengths of ultraviolet—those not eliminated by the filter—also caused the synthesis of organic compounds? This had not happened with any of the soils or wavelengths Horowitz and Hubbard had tested on Earth, but the chemistry of Mars had already exhibited such unusual properties, and the effects of ultraviolet light had been so little studied, that this possibility had to be considered. Another possibility was that hydrogen peroxide might be involved; though it was an oxidant, it also contained the hydrogen necessary for the kind of reductive, autotrophic reaction Horowitz's instrument was designed to detect. The reaction might then proceed in a manner involving both hydrogen peroxide and carbon monoxide, which he already knew created formic acid. Horowitz and others liked this possibility because it suggested that the results from all three instruments had something in common—oxidants. (The price, of course, was to reverse the biologists' conviction that Horowitz's instrument was not vulnerable to these chemicals.) Suspicion also fell on the pervasive red iron oxide—hematite or whatever it was. Oxides such as hematite are oxidants too, though weak ones; and although their oxygen is not necessarily released by water, as that of other oxidants is, perhaps oxygen might be released by other means, having to do with the ultraviolet light, to make one of the photolytic syntheses possible. Or perhaps something altogether different was going on. "If it turned out that on Mars even *visible* light caused something on the surface to synthesize organic compounds, that would be a remarkable finding," Klein said at the time. The scientists took no comfort in the fact that two of the likeliest culprits were ones that would have been known to any of the children Sagan had been addressing, before the launch of Viking 1, around the pool at the Ramada Inn: rust and a common disinfectant their mothers may also have used to whiten the wash. Clearly, Horowitz's instrument, though it had been designed with the Martian environment in mind, turned out, through nobody's fault, to have a potentially devastating Achilles' heel.

On sol 21—August 10 on Earth—several members of the biology team participated in a public symposium held in the von Karman Auditorium to discuss matters as they stood. Klein, expressing the general bewilderment, said: "Mars is really talking to us and telling us something; the question is whether Mars is talking with a forked tongue or giving us the straight dope."

The reporters present regarded Klein as a sort of barometer of what the biology team thought. Clearly, Horowitz's results had unsettled him as much as they had Horowitz, who once again sounded a note of caution in the face of the apparently positive evidence: "If we seem to be slightly evasive in our answers, it is because we are all terribly conscious of the fact that we might well be wrong in anything we say. And nobody wants to be wrong in public on a question as important as that of life on Mars." As chairman of the meeting back in 1969 at which the wishful scientist had announced the possibility of life on the planet, Horowitz, if he had known about the statement ahead of time, would have tried to head it off. Certainly nobody was going to make such an announcement today. Even so, one biochemist—Leslie Orgel, of the Salk Institute, who was one of the early researchers with the complex carbon polymer chains that contain or transmit the genetic code, DNA and RNA, and who was a member of the molecular analysis team, which worked with the GCMS—said he was now retracting an offer he had once made, when he was certain there was no life on Mars, to eat a sample of Martian soil if anyone should bring some back. It turned out, though, that Orgel was retracting his offer not because he thought microbes were any more likely but because of the possibility of oxidants; their reaction in water would give him a sore mouth.

One after another, all the scientists expressed the same note of uncertainty, causing Lederberg to remark that cautiousness was the one point of agreement among the scientists. As had been the case with Horowitz for some time, they all suddenly seemed aware of the possibility that if you design machines to find signs of life on Mars, that is what, irrespective of its actual presence, you will very likely find. Even Carl Sagan was relatively cautious. Although he wasn't present at the meeting, he would say in private a little later: "Two things seem to me of extreme importance about the results from Mars: that these instruments were designed to detect life, and that when they go to Mars they all return active signals. What does this mean? If it turns out that these signals are *not* biological, it means that there are nonbiological processes simulating life just lying around on the soil. It's easy to pick these processes up and evolve life with them. It means that there, sitting all around, are all the oxidizing and reducing chemistry it needs—little metabolic cycles of oxidation and reduction are going on all the time by themselves. Now the big event in the evolution of life is self-replication, the development of the genetic material in the oceans, and what I'm talking about has no bearing on that; but once these replicating things are splashed on land, they can then incorporate these kinds of energy-producing mechanisms. So even if the less interesting alternative is correct and the Viking results turn out to be chemical, they have deep implications for terrestrial life." Horowitz, impressed with Sagan's restraint, said he was worried about the astronomer's spirits. He needn't have been, for before long Sagan would be saying things Horowitz could take strong exception to once again.

Lederberg, whose remarks on Warburg had been stimulated by the same sort of evidence that had prompted Sagan's remarks and who was at the symposium, asked the other members of the panel a question that revealed the unsettled state they were all in. What could yet happen inside the instruments that would convince them that some phenomenon was of biological origin? Klein said he felt the surest sign of life could come from Oyama's instrument: if during its lengthy incubation, on which it was now embarking, some gases reappeared after having begun to disappear, or if there were any other unexplained fluctuations in the rates of gas production, these results would be very hard to explain chemically. "Although I'm sure my colleagues would come up with some very, very strange chemistry to try to explain them," Klein concluded. Others mentioned the possibility that Levin's carbon dioxide curve might yet take off exponentially or that the chemical explanation of Horowitz's count of 96 might be proved erroneous, leaving biology the sole contender. Lederberg, for his part, felt there was no evidence the three instruments could supply that would prove beyond doubt the presence of life on Mars. "For that, we have to see growth, we have to see reproduction, we have to see the possibility of evolution and diversification, and those are difficult experiments to do," he said. He wished Viking had a microscope.

Horowitz said he wouldn't be satisfied until he had some samples of Martian dirt and rocks in his own laboratory on Earth. "We also have Viking 2 to look forward to," Lederberg said. "The Viking 1 site is very dry, very arid, and obviously very hostile. I think many of us feel reinforced about the desirability of looking for a habitat somewhat different than the one we're in now." Lederberg, of course, was invoking the idea he and Sagan had developed many years before, that some areas of Mars might be more hospitable than others—indeed, there might be what the two scientists had called microenvironments on Mars, small oases where life might thrive.

Lederberg didn't go as far as Sagan, though, who was currently campaigning for an unmanned spacecraft mounted on wheels that could scout out such places. "I'm thinking of a vehicle that could wander hundreds to thousands of kilometers in a nominal lifetime," Sagan said on another occasion when he was rhapsodizing about a Martian rover. "It could wander to its own horizon every day. Anything you see today, you could examine close up tomorrow. You could have the capability of roaming over the polar ice. You could wander down a river where water once flowed. You could ask: Are there silts there? And organisms? But even if no organisms turned up, what a spectacular trip that would be!" In fact, engineers at J.P.L. had been working for some time on methods by which such a rover could "see" (either by means of a laser or by a combination of photography and computer) obstacles before it in order to steer around them without having to wait for a command from Earth for each turn of the wheel. And a committee of Viking scientists was already beginning to consider a future mission that would include not only an orbiter and a rover but also a third element, several hard-landing devices called penetrators, which would be released like darts as each lander-rover descended, impacting in widely dispersed parts of the planet, beyond the range of the rovers. The front of each, like a rocket's nose cone, would sink several meters into the ground, while the rear—the fins—would remain on the surface to transmit information, very likely about oxidants as well as other matters.

At the symposium, the biologists, despite their difficulty in interpreting the evidence, resisted the notion that they were participating in the "horror show" predicted by the geologists. They were now glad that Viking had skipped over such intermediate exploratory steps as atmospheric probes and hard landers, which might have told them more about the Martian environment. Though it would have been useful to know ahead of time that the rocks and soil might well be coated with oxidants, a geological investigation would not have been likely to discover the presence of peroxides or superoxides. "Geological scientists very seldom do an experiment that disturbs the environment," said Oyama, whose own experiment did so by adding water. He had, of course, been attempting to perturb organisms, but had apparently perturbed the soil instead. "There was no geology experiment that would have performed this kind of analysis," Oyama said. He clearly enjoyed beating the geologists at their own game; his instrument had very likely made the major geochemical finding of the entire mission. "So I think we did the right thing with Viking; if we had gone more slowly in steps, we might very well not have detected these things," he said. Most of the biologists agreed with him, including Horowitz, who for the time being was inclined to think kindly of Oyama because of the alacrity with which he had sacrificed his ideas about biology on Mars. "No geology experiment would have added water," Horowitz said, adding handsomely, "and I wouldn't have, either. It was pure serendipity! Mars is so unknown that almost anything you do reveals something. If we'd adopted a slower, more rational approach, I don't know that we'd know what we know now."

The consensus was that muddling in with Viking was about the best thing they could have done. Even the order in which the results from the various experiments had come back was serendipitous. "Had Horowitz's results come out first, without the foreknowledge of some of the odd chemistry on Mars, we might have been much more confused than we are now," one biologist said. Serendipity, of course, has always been a respected element in science, and it continued to be a major element in Viking—as though Mars were some sort of giant laboratory for proving that scientists are no better than anyone else at anticipating the unknown. Indeed, in the rush of space discoveries over the last two decades, the scientists themselves were surprised at how often their guesses about the planets they were seeing for the first time proved wrong and at how, whenever that happened, it wasn't because they had failed to pick the right alternative of several under consideration, but because the true situation turned out to be one they had never envisioned at all.

With the biology instruments returning such ambiguous results, it was more important than ever to determine whether there were organic compounds on Mars. The reason that, back on sol 8, the GCMS had apparently received no sample—or to be more precise, had apparently not received at least 1 cubic centimeter, the minimum amount its capacity gauge could register—had to do with the nature of the soil on Mars, the nature of the Viking lander, and especially, the manner in which the lander was controlled. After the sampler arm had dug the trench to get soil for the biology package, it had gone back to the identical spot to get soil for the GCMS. Klaus Biemann, a slender, anxious-looking organic chemist, who was the leader of the molecular analysis team, had been sitting in the control center watching some electronic monitoring screens; after the scoop had supposedly dug the soil and dumped it into his hopper, a code number was supposed to show that the minimum amount of soil was inside his instrument, but the number had failed to appear in the proper place on the monitor. A little later, when Biemann saw a photograph of the trench, it didn't take him long to guess what had happened. On Earth, where the maneuver had been simulated countless times, returning to the same trench had always worked, because the soil, being rather coarse-grained, had always caved in after the first digging, so there was enough for a second helping. On Mars, this had not happened; instead of caving in, the walls had remained absolutely vertical, as though they were made of concrete, and so it was easy to see how the scoop could have returned no soil on its second try. The Martian soil turned out to be so fine-grained that it was cohesive, and the cohesiveness was reinforced by the low atmospheric pressure. (The soil was even finer than soil from the Moon, very likely because of the constant knocking about in dust storms. Whatever larger, sand-sized particles were left on Mars appeared, in Viking orbiter photographs, to have settled into vast fields of dunes around the north pole—the result, most likely, of sifting by global winds.) The scoop had gone back to the same hole because it was useful to have the GCMS and the biology instruments test the same soil. Also, in the lander's automatic style of operation, the sampler arm was incapable of breaking new ground more than twice in the same sequence; and because the inorganic chemistry experiment—the X-ray fluorescence spectrometer or XRFS, which also analyzed the soil—needed far more than either the biology package or the GCMS, it was obviously preferable for the XRFS to get the soil from a new digging.

The lander had to act automatically because there was then a 38-minute delay in round-trip radio communications between Earth and Mars; a complicated maneuver like sample collecting, requiring a score of consecutive commands, each of which had to be verified, could take a long time if it was done from J.P.L. It would have taken at least 3 days to halt Viking, compose a new program, and send it to the lander's computer—a sophisticated one that to a limited extent could, after a fashion, make its own decisions, based on data from the instruments, as a human brain would act on information from the senses. To Biemann, the control room seemed maddeningly relaxed; since so much of the work was done aboard the lander, the control room lacked the sense of urgency usually associated with such places. When the sampler arm automatically dug the new trench (actually, it simply extended the old trench 4 inches) and returned successfully with a heaping scoopful of soil for the XRFS, Biemann privately cursed all robots. (Despite Biemann's understandable annoyance, the engineers and scientists were pleased with the robot's performance. In fact, Sagan, who had earlier equated the number of words stored in the spacecraft's computer to the vocabulary of an 8-year-old child, later compared the lander's intelligence and decision-making powers to a grasshopper's, something he meant as a compliment. The computer had the ability to make certain rudimentary decisions, such as the order in which it sent back information from its instruments—something the people on the ground who were receiving the data could never predict. Orgel, Biemann's teammate, said once: "I'd rather have Viking up there than a graduate student. It doesn't have emotional problems; it doesn't tell you where to dig or what to do.")

The sampler arm had gone out again on sol 14, to try once more for the GCMS. Once more, Biemann was observing the monitors in the control center, and once more no number appeared in the data frame. For the second time, it seemed, no sample had been delivered. "I'm on the nonreceiving end of the system," Biemann said. With $55 million tied up in the GCMS, more people than just Biemann were dismayed. The Viking project manager, James Martin, deployed a team of 50 engineers to look into the problem. There had been what the engineers called a *no-go* from the sampler arm—a signal indicating that something had happened to cause the arm to stop in the middle of an operation. Once the scoop was in the ground, the arm was supposed to retract about 6 inches, rise, and then retract again. This time, instead of retracting the second time, it had stopped. On sol 18, after considering the possible causes, the engineers sent a new set of commands. The boom was commanded to extend a few inches before again responding to the command to retract for the second time; it was shifted to a lower angle to put less strain on the motor; and the maneuver was done in the early afternoon, when the temperature at Chryse had risen from its early morning low of −122° F to −22° F. The sampler arm came unstuck. In the future, all trenches would be dug in the same manner, though the restriction of having to dig in the middle of the day would pose some problems for the biologists.

Ever since sol 8, when the boom had apparently failed to get a sample the first time, Biemann had had a niggling suspicion that perhaps, after all, the GCMS *had* received a small amount of soil. He needed only 0.1 cubic centimeter—a tenth the minimum amount that would register on his gauge—to get a result; conceivably, he had enough in his instrument without knowing it. On sol 8, the sampler arm had indeed stopped over his hopper and gone through the motions of dropping soil; if it hadn't, there would have been a no-go. Also, in a recent photograph from the lander it looked as if a few grains of soil were scattered around the hopper. "The only way I can know for sure if there is no soil inside is to go up to Mars and look," Biemann said; he felt as though he were playing blindman's buff with his instrument—a common complaint among the scientists. A wrong decision could be expensive: he had only three ovens for pyrolyzing soil (like Horowitz's instrument, his heated the sample to break down organic compounds), and one of the three was broken.

On sol 17, Biemann had taken a chance and gone ahead with his experiment. The first step was to heat the soil to a fairly low temperature (200° C), which was high enough not only to drive out any carbon dioxide or carbon monoxide from the atmosphere that had got mixed in with the soil but also to evaporate any small organic compounds, which are more volatile than large ones. His gas chromatograph detected only a small amount of carbon dioxide—leaving unanswered the question whether there was a sample present. So he awaited the results of the second heating—at a higher temperature—with trepidation. Heating the soil (if there was any) to 500° C would decompose larger organic compounds. These fragments would then pass in a stream through the gas chromatograph—a coiled tube that separates the molecules by size—and then into the mass spectrometer, which would tell what they were. Biemann carried out this procedure on sol 23. He now found water (as well as more carbon dioxide), satisfying him that he had soil.

While most scientists had believed before the mission that the chances of finding life on Mars were poor, they had believed that the chances of finding organic compounds were good and that therefore the GCMS would prove more useful than the biology package. Biemann and Klein had both put the odds of finding organic compounds at about 50 percent; they felt that if the chemical theory of the origin of life was correct—that organic compounds were bound to form wherever the conditions were right—organic compounds almost had to be there. Very early in the Earth's history, of course, the oceans were a rich organic soup; if that had ever been the case on Mars— perhaps not in oceans, but in ponds or even puddles— detectable traces should still exist. Before the mission, even the biologists had been sure that Biemann's instrument would provide more substantial data than their own. This seemed so likely that the GCMS was designed to do much more than the biology instruments, which were intended simply to find out if there were microbes on Mars, yes or no. Having found organic compounds, the GCMS would go on to the next step—to try to determine which ones they were. Consequently, the GCMS was by far the most complex of the instruments; that is why it had cost as much as all three biology instruments combined. Now that Viking had landed, the discovery that there were probably oxidants on Mars reduced the chances of finding organic compounds, but the discovery of nitrogen, together with Horowitz and Hubbard's earlier discovery that certain wavelengths of ultraviolet light could generate organic compounds, raised the chances. Clearly, there were processes on Mars that made such compounds and other processes that destroyed them. Although Biemann did not know at what point the two would balance out, he still had hopes that the balance would be on the positive side and the GCMS would come up with something. Its abilities were generally admired, and Horowitz, a hard man to please, paid the GCMS what was for him the ultimate compliment: he said that it and his own instrument were the only ones needed to settle the question of life on Mars. In his opinion, life could exist there only if both instruments gave a positive response. One of these criteria, a positive response from his own, had, unexpectedly, been met. The other—a positive response from Biemann's—had always looked like a much surer thing.

The GCMS discovered no Martian organic compounds whatever, even though Biemann's instrument could detect most of the compounds he was looking for to a level of a few parts per billion. The result was unchallengeable, at least with respect to possible instrument anomalies, for the experiment turned out to contain its own control. During a test when the GCMS was still on Earth, using soil known to be sterile, the instrument had detected small quantities of certain organic chemicals—methyl chloride and benzine—which were solvents that had been used in cleaning the instrument; it had been impossible to eliminate all trace of them. The same chemicals, in the same small amounts, showed up in the data from Mars—evidence that the instrument was working perfectly.

Klein had once said that if there proved to be no organic compounds on Mars, that fact would lead him to question the chemical theory of the origin of life, which had contributed so much to the biologists' interest in Mars. However, that response would ignore not only the complexity of the present situation but also the resourcefulness of the biologists. "Before the mission, we all expected to find organic compounds on Mars," said Leslie Orgel. "When we got there, we found the surface very highly oxidized. We were wrong for unimportant reasons. Suppose I say to you that you will surely find a fruit shop open on Oxford Circus. You go there on a Sunday, and find that none are open. Now you could tell me I was wrong, but it's not because there were no fruit shops on Oxford Circus that you didn't find one open." Orgel and others were suggesting that the two adverse circumstances—the presence of oxidants and the absence of organic compounds—canceled each other out: the absence did not have to mean that no organic compounds had been or were being produced on Mars; it merely showed that at present they were being destroyed by the oxidants as fast as they were being created. One might have expected to find a certain amount of organic material carried by meteorites to the Martian surface, and the fact that not even *these* had turned up on Mars provided a further indication that something was destroying them.

Biologists studying Mars have always had a way of bouncing back in the face of data that might seem to outsiders to amply excuse them for abandoning all hope. "If we'd found organics, of course, that would have pushed us toward an organic explanation for data from the biology instruments," Orgel said. "But the reverse, although it is discouraging, does not necessarily rule out biology." The logic behind this conclusion was that the GCMS, like the biology instruments, was designed to prove the presence of something, not its absence. Now a number of biologists and biochemists began thinking up ways for life to exist on Mars even though the GCMS had failed to find organic compounds. Perhaps organic compounds were present after all, but in quantities too small for the GCMS to detect. Or if there were no organic compounds at Chryse, there might be some at

other places—for example, at the site where the second lander
was due to touch down shortly.

Biemann, who had once said he could imagine organic
chemistry, and life, without nitrogen, now came up with a
model for a Martian organism that would fit the new set of facts.
"The idea is that life originated when conditions on Mars were
benign, and then it adapted to the slow change," Biemann said.
"In harsh conditions, what would be more efficient than
scavengers? They might live in a colony; whenever one died, it
would immediately be devoured as food. That way, there would
be too small an amount of organic material left lying around for
the GCMS to detect." This theory was feasible because
Biemann's instrument, like Horowitz's, had an Achilles' heel.
Paradoxically almost, the GCMS was not sensitive enough to
detect organic molecules in individual living microorganisms,
which in relation to the total bulk of the soil were infinitesimal.
It could detect only the far greater amounts of organic material
that resulted after millions of organisms had died—very likely
over many generations—so the amount in the soil had built up.
Though the GCMS could detect most important organic
compounds down to a few parts per billion in the soil, the
average microorganism contains mostly water and very little
organic material; there would have to be at least 1 million
scavengers per gram of soil for the GCMS to have a chance of
detecting their presence directly. If there were fewer than that,
and if they were eating their dead, the GCMS would not find
them and neither would Oyama's instrument. Horowitz's and
Levin's instruments could, however, because their radioactive
labeling technique enabled them to detect far smaller concen-
trations of living organisms; in other words, a negative on the
GCMS or a negative on Oyama's instrument was not inconsistent with a positive on the other instruments—the very results,
Biemann pointed out, that they were getting. This situation
caused many arguments later.

Horowitz was not impressed by Biemann's scavenger. A little later, at one of the scientists' frequent conferences, when Biemann was explaining how the scavenger would devour all trace of organics in the soil, Horowitz replied: "But there is nothing on Mars that says this is the case! You have to endow your bugs with special qualities to fit the facts!" A voice from the audience said to Horowitz: "Aren't *you* doing the same thing when you create a special chemistry—ultraviolet synthesis of organics—to explain away the results of your experiment?" Horowitz admitted that his own results were indeed most difficult to explain, but he was soon back on the offensive. "The major fact that prevents us from offering a biological explanation on any of these instruments is that there are *no organics* down to the part-per-billion level, and to maintain that there *are* organics you have to invent a special model," he said. "It's totally ad hoc." Calling a theory ad hoc was about the worst criticism Horowitz could think of. Levin and his associate Straat, however, were extremely taken with Biemann's scavenger; since a scavenger would have to be an animal-like heterotroph, it, or something like it, might be causing their spectacular carbon dioxide curve. "Don't forget, you're dealing with another planet," said Straat, who saw nothing ad hoc about the scavenger. "Just because it's different doesn't mean it's excluded." She admitted to a slight feeling of disappointment when the GCMS results were announced. "That was the first really tough problem to cope with," she said. "It certainly gave a negative spin to the ball."

The rest of the search for life unfolded like a game of chess, its object being to pin down the elusive agent on Mars. Aboard the Viking 1 lander, and later aboard the Viking 2 lander, each of the three biology instruments ran a series of experiments called cycles; sometimes the cycles were run simultaneously, and sometimes, when one or another of the instruments skipped a cycle or did a longer incubation, they ran independently. Between cycles, the pipes, valves, chambers, chromatographs, and radioactivity detectors were flushed out by helium, an inert gas, which came from a single small tank—one of the few pieces of equipment shared by the three instruments. Changes in procedure or protocol could be made from one cycle to the next: Horowitz could add minute amounts of water, and he could turn his light on or off; Levin and Oyama could use more or less nutrient. The search was eminently suited to computers, with their ability to run down various alternatives a number of steps away, and the scientists spent a lot of time planning their moves far ahead, drawing up alternative strategies that depended for each step on the results of the previous move. When these alternatives were plotted on charts, they made branching patterns that the scientists called *decision trees*.

Oyama's instrument did not lend itself to decision trees as well as the two others because it would continue to incubate the same soil for much longer periods, purging gases and adding new nutrients an average of every 20 days. Although in the opinion of some of his colleagues Oyama took what amounted to pride of ownership in the oxidants, he had been optimistic for so long about life on Mars that he could not bring himself to drop the idea altogether. The average temperature outside the lander was −50° C—and extreme cold, he knew, could cause microbes to metabolize slowly. Some soil from Antarctica that Horowitz had once given him which contained a small number of sluggish microbes had taken up to 200 days to generate enough gases for his instrument to detect the microbes' presence. He would incubate the soil on Mars for 200 days; if nothing showed up at the end of that time, he would be ready to call it quits.

Horowitz and Levin would have a livelier time. In the middle of August—about 3 weeks before Viking 2 landed—Levin's label-release instrument and Horowitz's pyrolytic-release instrument began their second cycles of experiments. These were the controls, in which the soil was sterilized by being heated at 160° C for 3 hours before incubation—a step necessary to prove and evaluate the results of the first cycle and to prove that the instruments were working properly. Since Oyama's oxygen peak was not considered biological, there was no need for him to do a heat-sterilized control as part of a biological protocol—he would instead keep on adding nutrients to the same sample. He would not, however, have the same check on his first results that the others were getting; that would have to wait until Viking 2 landed. The controls for the two other instruments were, in effect, reversals of their first cycles, for this time a negative result would be taken as a sign of life; if after sterilization there was no reaction, the responsible agent could be presumed to have been killed by the heat and to have therefore been alive. Traditionally, in biology, a negative on a control such as this is considered essential to the acceptance as a sign of life of a positive on a previous experiment, for 160° C is not enough to destroy most inorganic compounds.

For their controls, Levin and Horowitz had planned to use soil from the first digging that remained in the hopper from the biology package; they wanted the soil for the controls to be as much as possible like the soil used in the first cycle, for that soil was known to be active. This was just as well, for although the sampler arm was now back in commission after the no-go on sol 14, they were a little nervous about using it. Inside each instrument, an empty incubation chamber was rotated under the hopper; then each chamber, with the soil in it, was rotated to the place where it was heated. Vents, or valves in the pipes leading from the chambers, were opened for the gases to escape. After that, the vents were closed, Levin's nutrients were added and both Levin's and Horowitz's soil was allowed to incubate exactly as it had the first time. Both results—which were reported on sol 31, or August 20, exactly 1 month after the first landing and 2.5 weeks before the second—were negative. Horowitz's second-peak count was 15, the maximum that could still be called inactive. Levin, who, as before, had got his results first, found that the line on his graph made an initial spurt up to about 1,000 counts a minute over background (this was later attributed to a residue of radioactive gas from the first cycle) and then dropped to about 100. At a later press conference summarizing the biology team's results thus far, Klein said that the active result of Levin's first cycle and the destruction of the activity by the heat-sterilized second cycle "would be interpreted in terrestrial soils as indicative of biological activity." The same, of course, could be said of Horowitz's results. Klein added, though, that Oyama's evidence that there might be oxidants, together with Biemann's failure so far to discover any organic material, still prevented that conclusion from being drawn concerning Mars.

Decision trees work very well in programming computers to solve mathematical problems; they work less well in plotting strategy in chess games because living opponents are not always predictable; and as Viking would prove over and over again, they wouldn't work at all well in pinning down the unknown agent on Mars. As far back as sol 11, when Klein had talked about the possibility of oxidants, he had recalled that unlike most chemicals, including superoxides, most peroxides tend to break down over the 3-hour time of the heating at temperatures ranging up to 160° C—the temperature of the sterilized controls. This was especially true of hydrogen peroxide, the peroxide most likely to exist on Mars and the one Horowitz was beginning to examine to explain the organic synthesis in his own instrument. "The perplexing position we are in," Klein had said at that time, "is that the very act of heating the soil to sterilize it may drive out the oxidative power, and then the subsequent control test will show a reduction in activity that would mimic a sterilization"—that is, would once again mimic a biological result.

When Klein now brought up the possibility that the agent might be a peroxide, Levin, who still did not believe in the oxidants and wished the others would stop mentioning them, pointed out somewhat caustically that Klein and the others had previously been touting superoxides in an attempt to explain Levin's spectacular carbon dioxide curve. Superoxides—so unstable with water—do not degrade at temperatures as low as 160°. Evidently, they were not present, or they would have survived the sterilization to release their oxygen when the nutrient was added. Could it be, Levin asked, that Klein and the others, faced with a drop in activity, were shifting their argument to peroxides? They admitted they were. As it happened, Klein thought it possible that the superoxides were in the soil, as the force of Oyama's result had indicated, but that they were incapable of oxidizing Levin's nutrient. For all the steepness of Levin's carbon dioxide curve on his first cycle, which had originally led many scientists to equate it with Oyama's oxygen peak, it in fact represented a weaker, less rapid reaction than Oyama's and hence was more in keeping with the peroxides. This was consistent with a recent finding that Levin's nutrient was incapable of being oxidized directly by free molecular oxygen (O_2). Since water breaks down peroxides more slowly than superoxides, the thinking now was that the water in Levin's nutrient also acted as a solvent, providing a pathway between the nutrient and the peroxide, which oxidized it directly. Whatever the case, Levin obviously felt he had exposed an inconsistency in his opponents' argument; in the future, he would make it his business to see if he could eliminate the peroxides as well as the superoxides, as he insisted he had just done. Levin's attempts to cast doubt on one after another of the chemical arguments helped form the pattern of the coming months. Other scientists, of course, would resist the attempts.

As for Horowitz, he had to admit that he was getting an even more definite sign of life than Levin; since his instrument was not designed to be sensitive to an oxidative reaction, he could not claim that a destruction of peroxides was responsible for his lifelike negative—just as earlier he had not been able to blame the oxidants for his lifelike positive. Furthermore, his negative count had destroyed the possibility that the filter in his xenon lamp was broken and had allowed the shorter wavelengths of ultraviolet light to come through and generate organic material in the soil; the lamp could not be broken, or the result of the control experiment would have been another 96 or higher. If chemicals were mimicking life, they were doing an awfully good job of it.

Horowitz toyed with a couple of ways out. Perhaps the synthesis of organic compounds under longer wavelengths of ultraviolet light—the reaction he thought might be taking place in his instrument—could be destroyed by heat; that would provide a chemical explanation for a negative. This was a shaky argument, and Horowitz knew it, for not only had he failed to identify this reaction, but the only reaction like it—the one involving shorter wavelengths originally discovered in tests several years earlier—was *not* vulnerable to heat. Another possibility was that the count of 15 he had just received was not a negative response after all. The criterion had been deliberately inflated to guard against a false positive, but in a control experiment this safeguard would work in reverse. A true negative was 3 counts. Accordingly, Horowitz wondered whether the present count of 15 might represent a small amount of activity; after the sterilization, of course, any positive response would be proof that the activity was chemical.

Whatever the merits of these arguments, they meant that for the time being, at least, he didn't have to be the one to verify the presence of life on Mars.

Horowitz was beginning to feel a little like a yo-yo. Carrying on the account of the fluctuations in the chances of life on Mars that had continued from the time of Lowell up through the landing of Viking 1, Horowitz said: "Then Oyama finds what look like lethal oxidants, and the hopes go down once more. Next, Levin gets his curve, I get my count of ninety-six, and the odds go up again. Those results are strongly undercut by the discovery of no organic chemicals, and down they go once more. And now Levin and I get low results on our controls that could confirm biological activity. And so on—up and down, up and down. Periodically someone finds something that puts the chances up or down by an order of magnitude. And it's happening faster and faster."

Levin and Horowitz got their third cycles under way on sol 36, or August 25, 9 days before Viking 2 touched down. The aim this time was to see if they could duplicate the results of the first cycles, for it is another rule among biologists never to accept a result as a sign of life unless it can be repeated. There was still some soil left in the hopper, but Levin and Horowitz now wanted fresh soil because in the 28 days since the original scoopful was dug out on sol 8 any organisms in the hopper soil might have died or the soil's chemistry might have been altered by being inside the spacecraft; in either case, the old soil would not be as reliable for duplicating the original experiment as a fresh sample from approximately the same place. Mindful of the sampler arm's no-go on sol 14, however, the biologists didn't want to throw out the soil in the hopper until they had the fresh sample in the scoop and the scoop poised overhead.

But according to the digging sequence, which had already been programmed into the lander computer's memory, the soil in the hopper had to be dumped before the digging could start.

When the biology team wasn't debating the latest evidence about life on Mars, erecting new decision trees, or planning for contingencies, it spent its time cooking up strategies to outwit the instruments. In order to hang on to the soil in the hopper until the last minute—or "in order to eat our cake and have it, too," Klein said afterward, recollecting the event unhappily— the engineers had to turn off all three biology instruments and then turn them back on, a complicated bit of programming. Everything went perfectly—except that some coolers in Horowitz's instrument did not get turned back on; the command to turn them on had inadvertently been omitted from the program the engineers had linked up to the computer. As a result, Horowitz's soil was incubated at a higher temperature than it should have been for 2.5 days—the time it took the engineering staff to notice the error and make the correction.

Still, there was no reason Horowitz could think of that he shouldn't be able to duplicate the 96 of his first cycle— especially since he had recalled a promising theory first proposed during pre-mission tests by his colleague, Jerry Hubbard, to explain that result chemically. This idea, which he and Hubbard called the exchange theory, avoided the difficulty of synthesizing organic compounds inside the box (a difficulty because the light was filtered) by having the organics synthesized *outside* it. If the organic material (in quantities too small for the GCMS to detect) was synthesized in the soil while the soil was still on the ground, by the unfiltered light of the Sun, then it was possible that after the soil was scooped into Horowitz's instrument it took on some of the radioactive carbon from his atmosphere—that under the influence of the filtered xenon light the radioactive carbon would change places with some of the carbon in the organic material. Although Hubbard had not seen this result in tests with terrestrial soils, he and Horowitz suspected that it might occur with the material on Mars, which appeared to be so chemically active, and that it would give a highly positive result. But when the result of Horowitz's third cycle at Chryse came back, the second peak was only 27—well within the range of a positive response, but a long way from duplicating the 96.

Immediately, Horowitz and Klein turned their attention to the matter of the temperature inside the incubation chamber, which instead of being 17° C, as it had been during Horowitz's first cycle, had gone up to 26° C during the first couple of days of the third cycle. Possibly, if the higher temperature was taken into consideration, the instrument might be considered to have duplicated its earlier result, for whatever was causing the reaction was evidently sensitive to heat. This they already knew, in gross terms, from the 160° sterilized control. Both Klein and Horowitz (and also Hubbard, who made the announcement to the biology team as Horowitz had a cold) felt that the present result—accidental as it was—was even more consistent with the presence of life than another 96 would have been because biological systems are far more sensitive than chemical systems to small changes in temperature. "When temperatures get warmer than an organism is used to, its proteins can denature. This means its enzymes are gone, its membrane can't operate, and it dies," Horowitz said later.

Horowitz and Klein were not planning any imminent announcement that there was life on Mars, however. The very fact that *all* the experiments were being conducted not at just 9° more than what Martian bugs would be used to, but at an average of some 60° higher—in part, so that Oyama's and Levin's liquid nutrients and the water in Horowitz's experiment wouldn't freeze—was one of the main reasons Horowitz had such strong feelings that none of the results were biological. This argument, of course, was not decisive, for much might depend on which 60° and which 9°—perhaps the 9° difference was the more critical. Or perhaps a chemical would turn up that was equally sensitive to those 9°. "The temperature doesn't prove anything," Klein said. "All we can say for sure is that we did not duplicate the original experiment—either its conditions or its results." Horowitz's results seemed to be taking him farther and farther in a direction he felt was wrong.

Levin, of course, was delighted with the results of the third cycle—not only Horowitz's but his own. Though the soil Levin used, like that used by Horowitz, was warmer to start with, because it had been acquired in the middle of the day in accordance with the new rules, Levin's instrument did not have the additional problem of the coolers being off, so conditions in his incubation chamber more nearly approximated those of his first cycle. After his first injection, the radioactivity count shot up rapidly, reaching about 10,000 after the second day, and then leveling off slowly as it moved up toward 16,000 over the background level—but not all the way up to 16,000, so the question of whether the active agent was attacking just the formate, as would be the case with a finicky organism, was still up in the air. Levin not only had duplicated his first result, thereby substantiating his spectacular curve and (in his opinion, at least) enhancing the possibility of life on Mars, but had exceeded the original result by 50 percent. Some biologists were perplexed that the greater activity went along with the some-what greater heat of the soil—the reverse of Horowitz's results. The discrepancy didn't bother Levin because the two instruments were looking for different types of reactions. He and Straat decided to let the incubation continue for a long time to see whether the curve, which on the first cycle had shown tantalizing signs of rising slowly just before the incubation was terminated, might take off exponentially—the steep upward swoop indicative of rapid reproduction of organisms. The long incubation would become a source of further irritation between Levin and Horowitz, for among other reasons, as long as both Levin and Oyama continued to incubate, Horowitz was fretful. He was eager to get through his cycles while the instruments were working so well. To delay, he felt, would be tempting fate.

A Residue of Doubt

Meanwhile, on August 7, 1976, Viking 2 arrived in orbit around Mars, and the scientists and engineers had even more trouble finding a safe spot for its lander to come down than they had had finding one for Viking 1's lander. A landing site, of course, had long ago been picked from photographs taken by Mariner 9: it was a spot in an area called Cydonia, near the Martian forty-fifth parallel of north latitude, about 1,000 miles northeast of Chryse Planitia, the Viking 1 site. However, the Viking biologists and biochemists had for some time been lobbying the project manager, James Martin, to move the Viking 2 landing site even farther north, to the fifty-fifth parallel, because there seemed to be a greater chance of finding water nearer the summer pole, which at the time was the northern one. Scientists had previously suggested that Mars could have an underground layer of permafrost, which would thicken and rise closer to the surface as it approached the poles, where it would burst through the surface to form the polar caps;

indeed, the Viking orbiters had discovered several craters that were surrounded by what looked like radiating, lobular mud flows—as though the heat of the impacts that made them had melted subterranean ice. Before Viking 1 landed, a scientist had suggested that vapor from the permafrost—or perhaps from a seasonal upper layer he called *tempafrost*—passed into the atmosphere with the coming of spring and solidified again in the fall; this, of course, would happen more readily in northern latitudes, where the ice was nearer the surface. The water vapor the Viking 1 orbiter had detected in the atmosphere increased toward the northern polar cap; consequently, the biologists began pushing harder than ever for a site near the fifty-fifth parallel. It was unwise to go north of that because of the risk to the lander of colder temperatures in winter, now less than 8 months off. (The lander's lubricants could freeze, a danger if the craft was exposed to temperatures much below $-50°$ C). Elsewhere, the biologists were encouraged by the sight of wispy white clouds—summertime phenomena—which proved to contain water; such clouds were common in low-lying canyon areas, mostly near the equator, where the lander could not touch down, and also in the apparently flatter north, where it could. Though the clouds were hardly the stuff of which deluges are made, they could at least provide a bit of moisture on the surface, for—like mist on Earth—they evidently settled on the ground at night when the air cooled and rose again in the morning when it warmed up: pairs of Viking 2 orbital photographs of the same area taken a few minutes apart showed clouds in the second photograph where there had been none in the first, and the clouds must have come out of the ground. The presence of ground mist was later confirmed by the lander's cameras, for an astronomer tracking Phobos, one of Mars's two moons, as it moved slowly toward the horizon discovered that it dimmed at a faster rate than it would if dust particles in the atmosphere were the only obstruction. (The two moons were the only heavenly bodies, in addition to the Sun, that were photographed by the lander; the optical system was not sensitive enough to take pictures of the stars or, in particular, of the Earth, as Sagan had once hoped.) This daily laying down and rising up of the mist would, of course, be minimal, affecting only the top millimeter or so of the soil, but that was enough to increase the biologists' hopes.

It was just as well that the biologists wanted to change landing sites, for when the scientists and engineers got a good look at Viking 2's orbital photographs, so much more comprehensive than Mariner's, they did not like what they saw of Cydonia. Craters stuck up out of the plain like knobs—evidently the meteoric impacts that made them had hardened the ground under them, and the wind had blown away the dust around them. They appeared dangerous to the Viking 2 lander. In addition, the terrain—gentle and sandy-looking in the Mariner imagery—now proved to be covered with boulders and, even worse, to be crisscrossed by a network of enormous cracks— reminiscent of cracks on terrain on Earth underlain by permafrost—which cut the terrain into vast polygons, like tiles on a bathroom floor.

Viking 2 was sent on a scouting expedition to search for a better place to land—a place, it was now decided, that would be in a higher latitude as well. Both Vikings were in highly elliptical orbits around the planet, flying far out to space, where their communications with Earth were not interfered with, and then zooming in toward Mars. The Viking 1 orbiter came in low over Chryse every time so it could relay messages and information between the lander and Earth; it orbited Mars in 1 sol precisely so that its low point was always over the same spot. Viking 2's orbit, however, was now changed so that it circled Mars a little more slowly, and as a result its low point each time was a little farther east than it had been; in this fashion its periapsis (its low point) would wander (or *walk*, as the engineers put it) around the planet once every 9 days. Later, after both landers were on the ground, the orbit of one satellite would be shifted to pass over both, freeing the other satellite to walk around the globe permanently; *this* craft's orbit would eventually be changed so that it passed closer to the poles.

After a lengthy search, a spot was chosen in the vast, flat, seemingly smooth area called Utopia. ("That's gotta be the place!" one member of Masursky's site certification team said when Utopia was first proposed.) It, too, proved to have its share of knobby craters and polygonal trenches, but Viking 2 touched down safely, nevertheless, on September 3—as it happened, on a small polygonal island, for the trenches came in small sizes, too. The site was almost exactly halfway around the planet— about 4,000 miles—from the first landing place at Chryse, and was about 25 degrees farther north, or almost on the fiftieth parallel, which was close to where the biologists wanted it. When the first pictures came back from the lander's cameras, however, the ground turned out to be littered with even more boulders than at Chryse; one of the lander's antennas had been dented, presumably by hitting a boulder. One footpad had apparently landed on a rock, for the craft was tilted about 8°. It had come down on part of the vast rocky ejecta plain surrounding a huge crater, Mie, whose center was a couple of hundred kilometers to the east. If possible, the second site looked even more barren than the first. There was hardly any variety. The horizon proved virtually flat, and the same rocky rubble extended everywhere; there wasn't even a sand dune to be seen, as there was at Chryse. Not only did the rocks look alike, but the ground was a uniform sort of hardpan that reminded many of the geologists of a type common in the Southwest called *caliche*—a clay that is solidified into a crust by the evaporation of water. The soil was as red as that at Chryse.

Because the two landers operated on the same radio frequency, Viking 1 was now used less; it would be monitored from time to time to keep tabs on continuing incubations and to get back an occasional picture. Though Lander 2 was identical to Lander 1, there were some changes in the digging sequence as a result of the sampler arm's no-go. And although the sample for the biology instrument would once again be dug on the eighth sol after the landing, the sample for the GCMS would not be collected immediately afterward; to ensure getting enough soil, this time the GCMS sample would be dug on the following sol, the second lander's ninth, when new ground was broken.

This meant that, for the initial sample, the GCMS had swapped places with the X-ray fluorescence spectrometer, or XRFS, which was the inorganic chemistry instrument. The XRFS team was delighted to make the switch: it now wanted to analyze some pebbles, of which there appeared to be many right where the biology sample would be taken; and the best way to get them would be to take what was left over after the soil for the biology instruments had been shaken through the sieve at the front of the digging scoop on sol 8. At Chryse, the XRFS had analyzed only dust, which, like the soil on the Moon, had turned out to be composed principally of iron, calcium, silicon, titanium, and aluminum—and since the dust was a mixture of material blown by the wind all over the planet, it was presumably the same at Utopia as at Chryse. Pebbles, being local, should be different. Once again, though, things went wrong. As the scoop was trying to dump the pebbles into the XRFS hopper on sol 8, there was a no-go. As a result, Biemann was unable to get his soil the next day. When the engineers eventually managed to fix the scoop, which delivered the XRFS its sample 5 days later, on sol 13, there were no pebbles; what the scientists had assumed were pebbles turned out to be compacted dust, which crumbled.

Meanwhile, Biemann and the rest of his team had decided that they didn't like the looks of the spot to the left of the lander, where the GCMS sample was supposed to come from; it seemed suspiciously like the soil from the first landing site, which had proved to be without organic compounds. To the right of the lander they saw what they thought was a more promising spot—one that had been nicknamed the Bonneville Salt Flats because the caliche there reminded the geologists of that area of the Utah desert. (Because they were skeptical of the idea of life on Mars, they liked to name patches of ground in front of the lander after the more lifeless spots on Earth, possibly to annoy the biologists.) To the organic chemists the caliche at least had the virtue that it was not known to be without organic chemicals. At first, the engineers resisted the last-minute switch, but in the end they went along with it because finding organic compounds now had the highest priority on the mission. The GCMS got its sample on sol 21, almost 2 weeks late; once again, the molecular analysis team would be the last to receive its information.

Before Oyama did his first cycle at Utopia with his gas-exchange instrument, he had to run some tests to verify whether the machine was working properly. Because the result he had received from the first lander was not thought to be biological, he had not run the kind of control experiments that were run with the two other biology instruments. Consequently, a few of the biologists had begun to question the oxygen peak from Viking 1—a result in which Oyama took great pride. Confirmation was important because the peak provided the main evidence for the theory that the surface of Mars was coated with oxidants, and the presence or absence of oxidants was crucial to the evaluation of the data of one, and perhaps both, of the other instruments. Therefore, on sol 8, when Oyama received his first soil sample, he held it for a whole day in the chamber to see whether the warmer temperature affected it at all. In three analyses that day, he got no oxygen. This test proved to Oyama's satisfaction that his instrument at Utopia (and by extension, the one at Chryse) was not subject to certain errors it had been accused of. Then, a day later, Oyama injected his nutrient into the soil from below so that water vapor came in contact with the soil, as had been done in the analogous occasion at Chryse; and, as before, the oxygen peak appeared. It was only about a quarter as high as the one at Chryse, but this was consistent with the theory that there was more moisture at Utopia: the greater the amount of moisture in the soil, the smaller the amount of oxidant. Even so, it was a substantial peak—more than enough to confirm the original one at Chryse. Oyama was clearly relieved.

Since Oyama still couldn't bring himself to give up altogether his former hopes for life on Mars, he intended to use his instrument at Utopia exactly as he had used the one at Chryse—for lengthy incubations—on the theory that microbes in cold climates would take longer to generate the telltale gases. Horowitz, who had earlier complimented Oyama on his willingness to give up the idea, now accused him of backsliding. Oyama's ambivalence annoyed Horowitz, despite the fact that Horowitz was in an ambivalent position himself—a position the exact reverse of Oyama's. He began referring to Oyama's hope of detecting a telltale gas—methane, for example, from a methanogen, which on Earth is a distant collateral ancestor of plant and animal life—as Oyama's Great Event. In Horowitz's opinion, Oyama's instrument used so much water that if a Great Event was discovered on, say, the hundredth day of his long incubation, Horowitz was all set to argue that it betokened only a terrestrial organism that had stowed away aboard Viking before the launch, bioshield or no bioshield. Horowitz thought that Oyama should divert his instrument on the Viking 2 lander to a purpose he felt it was more needed for—a chemical laboratory to study the oxidants. "If Oyama really wanted to nail down the oxidants, he could do it by modifying his experiment

on Viking 2, but he is too absorbed in the idea of life,"
Horowitz said several weeks later, when Oyama had not shown
any sign of following his advice. "He is *still* waiting for his Great
Event."

At the time Oyama was getting the results of his first cycle at
Utopia, Levin was getting the results of his. Levin's experiment
at Utopia had been carried out in the same manner as his first
and third experiments at Chryse, because one purpose was to
see how similar the soils from the two sites were. They proved
very similar: there was the same curve shooting upward,
reaching a count of about 11,000 after 2 sols (this was much
faster than with fairly rich terrestrial soils); continuing up to a
count on the radioactivity detector of around 14,000 on the
seventh sol from injection; and then there was a drop. When
Levin superimposed the curves from his first cycle at Chryse
and his first cycle at Utopia (and also the first few days of his
third cycle at Chryse, which was still going on for a long
incubation), they matched very well. The Utopia curve,
though, was even steeper and higher than the others; in the first
27 hours, it shot up about 25 percent farther than the first and
third cycles at Chryse. Still, the Utopia curve had not gone
above the level of the radioactivity that any one of the substrates
(the formate in particular) could provide.

The high degree of activity Levin found at Utopia was not
what could have been expected if oxidants were the cause of his
results as well as of Oyama's, whose oxygen peak was lower at
Utopia, and Levin was quick to seize on the discrepancy. Levin
always seemed more willing than the others to make the most of
any advantage; perhaps being hot on the scent of life, as he felt
he was, made him all the more tenacious. At a press conference
on September 16 (sol 12 for the second lander), he reminded his
audience that his control experiment at Chryse (his second

cycle there) had already imposed constraints on the chemical theory by apparently ruling out superoxides, the explanation then favored by Klein and Oyama. (When scientists argue, they love to impose what they call *constraints* on each others' theories.) Though Levin had not yet managed to rule out the peroxides too, he clearly felt he had done something almost as good. "We now find that a new constraint has been imposed on the chemical theory as a result of the fresh data at Utopia," he said, warming to his work. "We have some confirmation now that when water vapor is placed in the presence of the sample, oxygen comes off. We are at a site which is about ten times as rich in water vapor as the first site. . . . So on the basis of these two thoughts we might have expected to see less of a positive response in our experiment. Instead, we see a response that's about twenty-five percent higher than the response at the first site. So the new constraint is that if we are dealing with an oxidant it is likely that it will take more than one oxidant to perform the results seen in Oyama's experiment and the results seen in ours." When scientists argue, they not only like to impose constraints on each other, they also like to invoke Occam's razor—the rule that simple explanations should be favored over complex ones—and Levin clearly seemed to feel that two oxidants, although possible, made the chemical theory more remote.

Though not even Levin would say he thought life on Mars to be likely, the chances for life were clearly on the upswing, despite the apparent presence of oxidants and the apparent absence of organic compounds: both Levin's and Horowitz's instruments at Chryse had initially shown signs of activity; the controls at Chryse in both cases had seemed compatible with a biological interpretation; and in Levin's case the third cycle at Chryse had confirmed his first cycle, providing the duplication scientists require of an experiment if it is to be considered reliable; it could even be argued that Horowitz's third cycle at

189 Klein, Horowitz, Levin, and Oyama

Chryse had done the same, if the extra warmth resulting from the inoperative coolers was discounted. And now, at Utopia, Levin had just duplicated his original result again. Moreover, he was still looking forward to the further results from his third cycle at Chryse, where the long incubation might at any moment cause his curve to take off in an exponential rise that might indicate life. And of course, there were many months yet in which a Great Event might occur in Oyama's long incubation. All these circumstances, taken together with the difficulties the chemical arguments were encountering, heightened the feeling that this was an advent period for Martian biology. When the question was put to Klein at the September 16 press conference, however, he replied that he did not think it was. "I don't think we're that much further along," he said. He looked gray and tired; he had lost some of the sprightliness that had characterized his responses earlier. Increasingly, Klein was finding himself in the same predicament as Horowitz: feeling that the ambiguous evidence received from Mars was not biological but being unable to prove it.

Levin, of course, differed. In answer to the same question, he said: "I would say there certainly *is* progress . . . in that none of these steps has eliminated or diminished the possibility of a biological response." Joshua Lederberg had once credited Levin with taking on the scientifically useful role of devil's advocate: "It's very necessary that someone hold open the position that the active signals may be biological," he had said. Many of Levin's colleagues felt, though, that he was not just playing a role.

"Oyama's getting less activity while Levin gets more at Utopia is a good point for Levin to make, but it's not insurmountable," Horowitz said later. (The very fact that he regarded it as a point to be surmounted was another indication, if one was needed, of where his feelings lay; Sagan's argument against Bruce Murray's—that some scientists were disposed *not* to find life on Mars—was not, perhaps, totally without basis.) Because Oyama's instrument was many times more sensitive to the presence of oxygen than Levin's, which had to go through the additional step of inferring the oxygen from the carbon dioxide, the fact that somewhat more water could be inferred from Levin's instrument at Utopia than from Oyama's might not mean much, Horowitz argued, for with the additional step its margin of error might be too great. And, he continued, the presence in Oyama's nutrient of ascorbic acid, which absorbed oxygen, made the results of the two instruments difficult to compare.

Levin dismissed Horowitz's criticism as too conjectural to worry about. He was already planning his next cycle at Utopia, which he hoped would do more to resolve the argument. Since his most recent experiment had proved that the soil at the two sites was fairly similar, he chose not to do the control experiment he had done at Chryse, in which he had sterilized the soil for 3 hours at 160° C; he could assume that the results would be the same here—a cessation of activity—and so would not provide him with new information. This time, he planned to heat the soil to only 50° C before incubation—to do what he called a *cold sterilization*. If the activity ceased then or lessened significantly, it would almost have to be biological, since a chemical reaction would most likely not be affected by that small increase in temperature. At the very least, such a result would narrow the likelihood of a chemical explanation even further.

In the meantime, the results of Horowitz's first cycle at Utopia were coming in. Horowitz seemed plagued with bad luck. For one thing, his Utopia instrument had developed a leak in the pipe leading from the organic vapor trap to the radioactivity detector; 30 percent of the gas in the incubation chamber would be lost over the 5-day incubation period. Even though Horowitz knew he could compensate for the leak in analyzing his data, he hated to think what it boded for the future. Of more immediate concern was the fact that high temperatures had once again interfered with his plans, as they had during his third cycle at Chryse, when he was trying to duplicate his original 96. Utopia, being farther north than Chryse, was warmer, because in summer the Sun stayed in the sky longer there, and owing to the thinness of the atmosphere, this made a greater difference on Mars than it does on Earth. For his first cycle at Utopia, Horowitz had intended to try once more to duplicate the 96—not only to confirm that result but also to find out whether the soil at Utopia reacted the same way in his instrument as the soil at Chryse; like Levin, he wanted to establish what scientists call a baseline between the soils from the two places. During the original Chryse experiments, the lamp in the incubation chamber had been on. However, the engineers now told Horowitz he would be wise to turn it off, because the entire biology package, which Horowitz thought was too warm to begin with in relation to normal Martian temperatures, was getting still warmer. The inability to establish a baseline with Chryse would not matter, for Oyama and Levin had already done this—or so Horowitz thought. He regretted more the loss of another chance to duplicate the 96; if he failed to do so, that critical result would be open to challenge. (Horowitz was becoming increasingly attached to the 96, despite the inconvenience it caused him.) However, the Viking scientists always seemed able to reap advantage from adversity, if only because any unexpected situation provided a new set of circumstances to test; as one biologist said, almost anything they did on Mars would be interesting because it had never been done before. Running the experiment in the dark would eliminate the possibility of photosynthesis as an explanation for any positive result, though it would still leave open the possibility of dark fixation—the taking on of atmospheric carbon without light. If he got a positive reaction, it would mean something interesting was going on.

A positive second peak was what Horowitz got, although it was a very weak count—23. Possibly the instrument was broken. Or possibly he was detecting some sort of organism that could fix atmospheric carbon in the dark. Horowitz was not inclined to think he had found a dark-loving organism, but he couldn't prove he hadn't. His and Hubbard's new exchange theory—that organic compounds synthesized on the Martian surface had been scooped into his instrument, later taking on radioactive carbon from its atmosphere—was no help now, for Horowitz was sure that his filtered ultraviolet light was required for the radioactive carbon to change places with the organic carbon already in the soil.

Nevertheless, the 23, weak as it was, was enough of a positive to make him think something was going on and to make him wonder if the low positives he had been getting in some of his experiments—the borderline 15 of the Chryse control cycle and the 27 of the third Chryse cycle, as well as the 23 he had just received at Utopia—all indicated that a low-grade reaction of some sort, whether chemical or biological, was taking place on Mars. Whatever it was, it was quite different from the more dramatic reaction—the 96—he had got when the light was on and the temperature was low. Occam's razor notwithstanding, Horowitz now began to think he was looking at two reactions: one, more active, that occurred in light at low temperatures, and the other, less active, that occurred in the dark at higher temperatures. Indeed, it might be the higher temperature that destroyed the great reaction, allowing the lesser one to be detected. On that basis, Horowitz began to speculate further that the key to the whole problem was water—water in tiny amounts, which was present at the lower temperatures but was largely driven off at the higher ones. Utopia, of course, was damper than Chryse anyway. This theory fitted in with his observation following his last cycle at Chryse—just before the second landing—that higher temperatures seemed to cause lower results; perhaps the near absence of water was the explanation there too. Certainly something with hydrogen in it was required for the reduction of carbon dioxide to organic matter—the reaction his instrument was designed to detect—and together with hydrogen peroxide, water in tiny amounts was the best candidate. Horowitz resolved that on his next attempt to duplicate the 96 he would turn his lamp on, since the light seemed to be a requirement for the greater reaction; he would make sure his temperature was low enough to keep water in the chamber; and he would add approximately three times as much water vapor as existed in the Utopia soil. (Though this was a lot for Horowitz, it was far less than the others used.) With midsummer over, temperatures at Utopia were beginning to drop, so the engineers now made no objection when Horowitz wanted to turn his lamp on.

As Levin and Horowitz moved on to the second cycle, which both regarded as critical, they needed new soil. (Oyama didn't because he was still doing his long incubation.) Horowitz and Levin were soon at loggerheads about where the scoopful of soil was to come from. The first sample, which Levin, and also Oyama, had found to be similar in most respects to the soil at Chryse, had been taken at the Utopia lander's far left; this time, several members of the biology team wanted the sampler arm to swing around to the far right—to the Bonneville Salt Flats, which had already attracted the attention of Biemann. Sagan, too, was talking up the calichelike Bonneville material because the inorganic chemistry experiment, the XRFS, had confirmed that there were present on Mars those elements that go into a certain iron-rich version of a mineral clay called *montmorillonite*—something that had been proposed by Sagan, Pollack, and a colleague, Owen B. Toon, from the analysis of infrared spectra returned by Mariner 9. Sagan was interested because montmorillonite is a superb catalyst in a number of synthetic reactions in prebiological organic chemistry— including the combining of amino acids into long-chain molecules. (Subsequent work at NASA's Ames Research Center has shown that other metallic clays, such as nickel-rich and zinc-rich ones that might well also be present on a primitive planet, may have the same value; apparently the crystal lattices of the clays' molecular structure was just the size of the amino acid chains and tended to hold on to them so that additional molecules could add to them.) Of course, just because the ingredients were present didn't mean there *was* montmorillonite on Mars. Though Horowitz was less impressed with any resemblance between caliche and that clay, he too wanted some of the Bonneville soil simply because, whatever it was, it *looked* different from anything they had tested.

Levin, on the other hand, didn't want the calichelike soil, montmorillonite or no. He wanted the same type of soil he had used before because he knew it was active; since the purpose of the cold sterilization he was about to do was to see if a temperature as low as 50° C killed the response, he required soil that had given a positive reaction before. Almost all the biology team, which devoted several of its daily meetings to the dispute, favored moving on to Bonneville. Levin and his associate, Patricia Ann Straat, carried the day, however, for they could show that getting the same sort of material was more critical for them than getting different material was for Horowitz. Horowitz, grumbling, went along with the team's decision. "The thing we spend most of our time talking about is the management of the instruments," he complained. "We're in lockstep. If I want some fresh soil, Levin has to take some, even if he doesn't want it, because there's no place to put it—his cells are full. If I want to start a new experiment next week, I can't, because he is in the middle of a long incubation, say. I accepted these conditions, but I resent them. If there was a Viking 3, I wouldn't agree to go again under these circumstances." The biology team's meetings were getting increasingly argumentative; the ambiguity of the data did not bring out the best in anyone. One member said, "It's not a congenial team"; another said, "The meetings are often stormy; they're not conducive to mutual assistance." Though there were frequent flare-ups, there were few lengthy exchanges; according to a third biologist, they specialized in what he called "quick-and-dirties."

On sol 28 at Utopia—October 1 at the Jet Propulsion Laboratory—the scoop went out once again. In Levin's instrument, after he had sterilized the soil at 50° C, which in the view of many scientists was too low a temperature to interfere with most chemical reactions, there was an injection of nutrient, and then the incubation went on normally—the idea being that if the activity known from the first cycle to be in the soil was reduced or wiped out by that temperature it was very likely biological because this would suggest that there had been death or damaging of organisms. When Levin's results began to come back, the curve started to rise, though not as swiftly as it had in the past. However, when it reached a count of about 1,600, instead of continuing up, it dropped sharply back, remaining from then on at about the 1,000 level—a negligible amount, since it included the background level. The low result was much like the one produced by the earlier control at Chryse, the sterilization at 160°; evidently, the 50° temperature had killed the reaction almost as effectively. The trouble was that the line on Levin's graph, instead of being straight, as it had been with the earlier control, was wavy; after the first few sols, during which the wave pattern was relatively erratic, the waves seemed to occur regularly on each sol—possibly because of the rise and fall of the Martian temperature. Because there was no obvious explanation for this, it threw doubt on the validity of the cycle: perhaps the machine was broken. Klein, Horowitz, and others, who had good reason to be dismayed if the data were correct, suggested that Levin's instrument was leaking, in which case it might not have registered greater activity if there had been any. Levin and Straat, after troubleshooting signals had been sent up and back, announced that no instrument anomaly had been found. Still, no one could explain the wavy lines, and the results would be indeterminable until someone could.

Even so, the arguments started. On the assumption that the results were correct, Levin said at once that in his view the data were "mind boggling." He felt triumphant. He had placed what he considered a constraint on the chemical theory by showing in his first cycle at Utopia that two different oxidants had to be involved. He had imposed another constraint when, during his Chryse control cycle, the 160° heat had killed the activity, thereby presumably eliminating most of the superoxides (and a few of the peroxides as well). And now here was a third constraint. Since the 50° heat of his cold sterilization—which wouldn't destroy most chemicals—had evidently destroyed much of this one, Levin felt that he could at last rule out the rest of the peroxides too. In short, he and Straat felt that the chemical theory had been severely constrained. "If no one can prove that the activity is chemical, then no one can say it's not biological," said Straat, turning the biologists' own guideline on its head with as many double negatives as Sagan was accustomed to mustering. Although Levin and Straat were not prepared to say that they had actually discovered life on Mars, they pointed out that living systems—as Horowitz himself had indicated when he was in the same position earlier—were obviously the most likely to be wiped out by a rise in temperature to a mere 50° C, the equivalent of 122° F. Horowitz, however, said that, with all this talk of temperature changes being favorable to life, he wondered, as he had before, why Levin and Straat never worried about the difference—60°, on the average—between the temperature inside the instrument and the cold outside.

Though the scientists favoring chemistry felt that more and more ground was being cut out from under them, they were clearly not ready to capitulate. Klein admitted, with some double negatives of his own (the ambiguity of the data seemed to demand a convoluted diction), that "the opportunity to rule out the presence of biology in the sample did not materialize because we were not able at this point rigorously to rule out the presence of living systems as an explanation."

Klein, in fact, was less sure than Levin that all the peroxides had been eliminated by the cold sterilization. Back before the 160° heat-sterilized control at Chryse, he had suspected that the very act of heating the soil could drive out the oxidant so that the control would show a reduction in activity mimicking a biological result. The same might be true of the 50° sterilization; under certain circumstances, he believed, hydrogen peroxide could evaporate at such a low temperature, giving results that simulated a living system. Horowitz agreed; once again this oxidant was dovetailing with a theory for his own results. Levin, however, was asserting that hydrogen peroxide probably could not exist for any length of time on the surface of Mars because ultraviolet light, to which it was sensitive, would destroy it as fast as it could be produced. Klein and most others disagreed; pure hydrogen peroxide might be destroyed in this fashion, but on Mars, where it was not pure because it was mixed with the soil, this would not happen; the silicon in the soil is very good protection against ultraviolet light and would act as a shield. In other respects, though, not enough was

known about the behavior of oxidants under the low temperatures and pressures that existed on Mars because there had never
been any reason to study this before. For that reason, Klein was
not at all sure that Levin had succeeded in eliminating the
superoxides, either. Nor did he think it ridiculous to assume
that there might be more than one oxidant involved; if Mars had
a highly oxidizing environment, without living organisms to
control it, why shouldn't there be a variety of oxidants?

If Levin felt that something in the soil at Utopia had tugged
ambiguously at his fishing line, leaving waves on his graph,
Horowitz found, when he pulled in *his* line, that his worm was
gone. In contrast to Levin, whose results, even when they were
unexpected, always seemed to carry him forward, Horowitz's
were increasingly leaving him in the doldrums. One of his
associates, George Hobby, said: "Even though we plan carefully, with a decision tree to predetermine what to do next, we're
continually surprised when the results come in, and we are
forever having to make another tree." For the second cycle of
Horowitz's instrument at Utopia, the lamp was turned on, and a
small amount of water vapor was injected. He was confident
that this time he would achieve a high count. The result was a
count of 2. For Horowitz, it was another reversal: the first
indisputably negative result he had received—less than the 15
he had got on his heat-sterilized control in the second cycle at
Chryse, which could equally well be read as being at the lower
limit of active. Clearly, his ideas about those two reactions—the
drier being the less active, the wetter the more active—needed
revising. He was further perplexed by the fact that the first-peak
result of the latest cycle, which had to do with the atmosphere
in the chamber, was twice as high as normal, but he concluded
this was very likely attributable to the water vapor. He went on

to consider the possibility that his instrument, like Levin's, might be malfunctioning; indeed, the leakage in Horowitz's instrument at Utopia had sped up. Even so, he was confident that he had allowed for the leak in his results and that the count of 2 was correct. But, like Levin, he could not be absolutely certain that his instrument was working properly until the results from the next cycle came back.

On the assumption that the instrument had given accurate data, the only possible conclusion was the exact opposite of what Horowitz had expected: instead of accelerating his reaction, whatever it was, the addition of water utterly destroyed it. On top of everything else, he was amazed that the addition of relatively little water had had such a devastating effect. This, he knew, was important; he even thought he knew *why* it was important. He thought the result was chemical. One possible explanation was that the water had destroyed oxidants and released oxygen in his instruments, as it had presumably done in the other instruments. This had not happened to him before simply because he had never before added water vapor. The methods he had thought up earlier to explain his active count by the synthesis under his filtered light of organic chemicals, using hydrogen peroxide, would have been put out of action in this manner. Since there was no reason to think the filter was broken, most of his remaining theories wouldn't apply, either. There were two theories left that would avoid these conclusions. One was the exchange theory, in which the organic compounds had been formed outside the spacecraft, later taking on radioactive carbon by exchange from Horowitz's atmosphere; the other was the hematite theory, which he had thought of after the first Chryse result, in which hematite (or some other iron oxide that reddened the rocks and dust of Mars) catalyzed the synthesis of organic compounds under the filtered xenon lamp—that is, under the longer, visible wavelengths of ultraviolet. In experiments already under way at the Ames Research Center, the actual hematite process appeared to be immune to water. However, the water would act as a solvent that would provide a liquid pathway between the organics and the oxidants that would destroy them. The same would be true of any organics that resulted from the exchange process. Either theory, then, would explain his negative. In other words, the count of 2 and the destruction of the activity were consistent with all the chemical theories, and hence, Horowitz said, his result "smelled" to him like an overwhelmingly chemical one. The trouble was that the negative could just as well be an overwhelming biological result; unfortunately, the experiment had done nothing to rigorously exclude biology. The water could equally well have drowned the microorganisms.

Several of the scientists were beginning to believe that the question of life on Mars was less apt to be resolved by the instruments there than by work in laboratories on Earth. "I don't think there is any way you can prove some of the results are not biological from the data alone," Leslie Orgel, the member of the molecular analysis team who had recently reneged on a promise to eat Martian soil, said at a symposium in October, not long after the results of Horowitz's second cycle came in. "However, if in the laboratory we find a good chemical explanation for, say, Horowitz's results, people will lose interest in a biological one. Conversely, if, a year from now, no one comes up with a material which, under the filtered light, produces organics in Martian-type soil, then we're in trouble. But I predict that they will." (This was too much for Sagan, who was present and who interjected, "Yes, and when chemistry fails, we'll be forced into biology!"—a statement that turned the biologists' cautious ground rule into a formula for finding life on Mars.) Sometime afterward, Horowitz received preliminary word from the Ames Research Center that the process involving a form of hematite—and, very likely, carbon monoxide—under conditions like Horowitz's was producing low active results. These results—coming amid so much ambiguous evidence—heartened him.

Periodically, the scientists and engineers at J.P.L. tuned in on the data that were accumulating from the first lander. Levin's long incubation at Chryse, an extension of his third cycle, which had gone on for some 50 days and had held up Horowitz, had recently been terminated; the results were a disappointment to Levin and Straat. The curve had never taken off exponentially, as it would have if organisms had thrived and reproduced in the incubation chamber. And once the background level from the lander's radioisotopic generators had been subtracted, the count—although still slowly rising—had never gone above 15,000, leaving unsettled the question of whether more than one substrate in the nutrient—the formate, for instance—was being oxidized. Moreover, several biologists were now disenchanted with the notion that the counts under 16,000—those accommodated by any one of the substrates—suggested a finicky organism attacking one substrate. It was possible that on Mars a chemical reaction *would* take place with only one substrate. It was also possible that the count represented carbon oxidized in part from several substrates, and the fact that it totaled 15,000 was just a coincidence. Accordingly, Levin was deprived of even this possible biological loophole in the failure of his incubation to take off. When, 16 sols after the initial injection, the counts per minute had risen only from about 14,000 to a little over 15,000 (including background level), Levin and Straat had ordered a second injection. In the fashion that had become familiar, the curve had dropped back before resuming its slow climb; that climb had continued until the forty-second sol without topping the 15,000 mark. Then there was a third injection, another drop, and another slow climb, which continued until the experiment ended. Levin seemed gloomy. Though his attacks on the chemical argument were increasingly successful—in his opinion, at least—he regretted his inability to prove that his results were biological. The sluggish rise, he admitted, was not consistent with life— although, he added hastily, it wasn't inconsistent with it either. "You could say that the bugs were killed by the time of the second injection," Levin said, repeating an old argument. Horowitz felt the results were clearly nonbiological, and he said so in no uncertain terms. The long incubation had done nothing to improve anyone's temper. Klein was finding the team meetings increasingly difficult to chair as some of the members were less and less interested in listening to each other. Lederberg, who was more and more inclining toward a chemical explanation, had largely stopped attending biology team meetings; so had Alexander Rich of MIT. In Horowitz's opinion, these were the outstanding members of the team, and he regretted their absence.

Though Oyama's instrument at Chryse was still ticking away on its 200-day incubation, the termination of Levin's long incubation there meant that Horowitz could at last get new soil. He had been growing increasingly edgy, for already it was late October, and in a few weeks Mars would disappear behind the Sun. There would be a period of about 1 month when communications with the planet would be impossible, and no one could guarantee what condition the instruments would be in afterward. Indeed, Horowitz had a built-in leak in his Chryse instrument, which was possibly more serious than the one in the pipe leading to his detector in Utopia. In his instruments in both landers, his radioactive carbon dioxide gas was contained in steel bottles sealed by metal membranes. After the landings, the seals were punctured in a manner that caused slow, predictable leaks such that in the Chryse lander, which had been on Mars 7 weeks longer, 90 percent of the gas would be gone by the time the experiments began again. (The identical leak in the Utopia instrument lagged far behind as the Utopia lander had arrived so much more recently.) Horowitz was more anxious than the others to get on with things anyway because he had more to do than either of the others; he could carry out six cycles with each of his instruments—two more cycles than Levin could. (This was because he had three incubation chambers in each instrument and could use each chamber twice; the high temperatures he used during pyrolysis prevented the old soil from contaminating the new.) Horowitz's greater capacity increased his annoyance at having to synchronize his cycles with the others'. He particularly resented the fact that the others were preventing him from doing an experiment he had increasingly come to think was essential—an incubation at temperatures well below freezing, such as Martian organisms would normally exist under—because the other biologists' liquid nutrients (which he disapproved of anyway) would freeze. (Even though he was glad he had brought a little water himself, and had even used it and might again, he still did not regard it essential to his experiment.) He began urging that a cold incubation start in November, before Mars disappeared behind the Sun, regardless of the effect on the other instruments. He hoped to break out of the synchronized scheme for good. But Levin and Oyama voted him down.

This was a frustrating time for everyone. Levin, too, found he could not do the experiment he wanted—one he called the *double squirt.* Some other members of the biology team, who believed his results had been caused by chemistry, had been going around saying that his spectacular carbon dioxide curves had been caused not by a biological event but by the depletion of some substance either in the Martian soil or in his nutrient—a suggestion Levin did not readily embrace. Still, it was a loose end that had to be nailed down. Accordingly, he wanted, in his fourth cycle at Chryse, to do a double squirt on his first injection, the idea being that if the determining agent was in the nutrient his curve would go up twice as far. The experiment might also get at the question of whether more than one of his substrates was being attacked. However, Levin's own associates, including Straat, and the rest of the biology team, including Klein, would have none of Levin's double squirt—for the present, at least. Because of the anomalous undulations in the data from Levin's cold sterilization at Utopia, they all agreed that it was best to save his last remaining chamber at Chryse until he had found out whether his instrument at Utopia was working properly. If it was, Levin could always do the double squirt at Chryse later, but if it wasn't he would have to attempt the cold sterilization once more; that experiment was regarded as absolutely essential, and the only remaining incubation chamber would be at Chryse. Levin eventually capitulated, even though it meant accepting surplus soil in his Chryse instrument. That was something he had been trying to avoid, and it had been the reason Horowitz had been delayed so long.

Horowitz, with his cold incubation postponed and his leak at Chryse continuing, was in a hurry to move on, but he too had a revolution on his hands. He had wanted to carry out what he called a *sudden-death* experiment to see if he could rule out biology once and for all as an explanation for his own results so far. He planned to sterilize his soil at 160° (to kill all organisms); to use relatively large amounts of moisture (to further stamp out activity); and to turn his lamp on (to encourage the long-wave photochemistry he thought the likeliest explanation for his positive results). "That way, any positive response would be bound to be chemical," he said. The sudden death, though, was destined to join the double squirt and the cold incubation on the shelf. (All would be attempted later.) Klein and also Horowitz's own associates decided unanimously that the most important thing Horowitz could do with his fourth cycle at Chryse was to attempt once again to duplicate the original Chryse result of 96—something he had already attempted twice with no success. The incubation chamber that was about to be used at Chryse was Horowitz's last empty one there, and everyone considered an empty chamber more reliable than a used one for duplicating the original result.

The result, when it came back, 5 days later, was a count of 35—far short of the 96, but still well within the active range. Horowitz noted that the soil had once again been gathered in the middle of the day and was therefore hotter than it had been during the original incubation at Chryse—a point he had made the last time he tried to duplicate his original result there. Furthermore, the normal rate of leakage was such that only about one-third of the radioactive carbon dioxide gas was left, and consequently, lower peaks were to be expected. As he had on that earlier occasion, Horowitz suggested that a 35 now might well be the equivalent of a 96 earlier. He knew, however, that this interpretation was highly speculative.

As Klein looked over Horowitz's second-peak results from Chryse, he was puzzled:

Chryse 1—original result	96
Chryse 2—heat-sterilized control	15
Chryse 3—attempt to duplicate original	27
Chryse 4—attempt to duplicate original	35

They didn't seem to add up. If the heat caused the lower results, and did so by driving off the moisture, then one thing was going on at Chryse, where the warmer, drier experiments (Chryse 3 and 4) gave low results, and quite another thing at Utopia, where the experiments had been damper, and the results were even lower.

Both Levin and Horowitz now turned their attention from Chryse back to Utopia, for by going ahead with another cycle there they could at least determine whether their instruments were in good working order, thereby indirectly verifying the results of their questionable second cycles. And an opportunity had presented itself there that interested everyone. Back on sol 21 at Utopia, when the sampler arm acquired soil for the GCMS, the scientists had decided it would be useful if the inorganic compound analyzer, the XRFS, examined a sample of the same soil, and accordingly, on the twenty-ninth and thirtieth sols the sampler arm went into action again. This time, as it dug a trench adjacent to the one it had made a week before for the GCMS, the sampler pushed too far forward and nudged a rock, which moved. The possibility that the sampler arm might be able to move rocks had never seriously occurred to anyone, and as soon as it did a number of scientists thought it would be a good idea to get some soil from underneath a rock. Such soil would have been protected from ultraviolet light, which is lethal to organic compounds, and since ultraviolet light is essential to the formation of oxidants, perhaps there would be no oxidants down there. And there might be moisture. All in all, there was a greater chance of finding not only organic compounds but maybe even life in rock-shaded soil.

Everyone wanted some of that soil—even Oyama, who was in the middle of a long incubation at Utopia. He decided to terminate it, for the soil seemed depressingly like what he was still incubating at Chryse. And Biemann, who had not found any organic compounds in the Bonneville soil (a result that would have prevented the argument between Horowitz and Levin if they had known about it) and had begun to suspect that he would never find any organic compounds on Mars, felt hopeful once more. The idea of the subrock sample, as they called it, seemed to have come along at just the right time to take everyone's minds off Levin's and Oyama's continued failures to turn up signs of life in their long incubations and off Biemann's second result indicating a lack of organics. There was a feeling of anticipation around J.P.L. that had not been present since the first cycles after the first landing; it resulted, if not from a tug on the fishing line, then from the sighting of a promising new pool in which to cast. Sagan in particular took great interest; first with the Moon and later with Mars he had been partial to the idea of life's existing snugly underground, and for this and other reasons he had recently been circulating a memo among the Viking scientists trying to gain support for having a sampler dig a trench a foot deep to get subsurface soil.

With all the thoroughness NASA normally lavishes on such decisions, a group of geologists set about analyzing a number of candidate rocks to push, meticulously grading them on a variety of criteria such as their grippability, accessibility, sampleability, and rollability. The rock that scored the highest turned out to be the only one that couldn't be budged; instead, when the sampler arm shoved against it, the lander itself was pushed slightly out of place. On sol 37, after further trials, the sampler managed to nudge another rock aside far enough to take a sample from underneath it. This was allocated to the GCMS, which now had an even higher priority than the biology instruments. The biologists got their sample from beneath another rock on sol 51; their soil was scooped up and deposited in the hopper as quickly as possible after the rock was moved, to reduce the chance that it would dry out or be struck by ultraviolet light. (To reduce these risks even further, the biologists had requested that the digging be done at night—what Sagan, who was emphatically in favor of the idea, called a "dark dig"—but the engineers refused because of their reluctance to operate the scoop in the extreme nighttime cold. They finally relaxed the rules enough so that the job was done in the early morning, shortly after sunup.) All the instruments' experiments were run in the usual way, except that this time Horowitz left his lamp off, on the theory that organisms from under a rock would be more likely to react in the dark.

The results were something of a letdown. Once again, Biemann found no trace of organic chemicals. The soil in Oyama's instrument yielded the lowest oxygen peak he had got yet, but that result simply indicated the smallest amount of oxidants and the most water—the conditions the scientists had expected to find underground. Only Levin was able to find much hope for biology: once again, he got a high radioactive gas curve. If the chemical theory was correct, of course, his curve should have dropped in proportion to Oyama's peak—the same argument Levin had made after the first cycle at Utopia, when Oyama's oxygen peak dropped to a quarter of his original one, presumably because the extra moisture in the soil had curtailed the formation of oxidants. This time, although there should have been even more water in the soil, causing Oyama's peak to dwindle even further, Levin's results were almost the same as both his first and third at Chryse and his first at Utopia. Levin, who thought he had already shot down the oxidant theory, now felt he had it in the bag. Though he still couldn't produce an exponential curve, he planned to let his incubation of the subrock sample go on for a long, long time, and Oyama would do the same.

The best piece of news from Levin's point of view was an indirect one: his instrument at Utopia had worked so well during its third cycle that it had clearly not been leaking or malfunctioning earlier, and therefore the results of the second cycle—the cold sterilization at 50°—were very likely correct. In other words, despite the mysterious undulations in those data, the line was basically flat, indicating a negative response—a finding that in Levin's opinion not only ruled out the peroxides as a possible agent but also was consistent with biology. However, the engineers had found a way to explain the undulations: moisture from Levin's nutrient may have condensed onto a cooler at the top of the instrument and then dropped onto the soil below, where it destroyed oxidants and released oxygen; his nutrient was then oxidized, causing a brief rise in the radioactivity count. Since there had been some minor thermal problems with the instrument, it was more than usually susceptible to changes in the outside temperature, and this could explain why the ups and downs of Levin's count had occurred daily. Levin still intended to repeat the cold sterilization he had carried out at Chryse—something that had to be done if such an important result was to be accepted. "If we can confirm that fifty degrees kills the reaction, I feel it will be mind boggling," he said.

Levin was obviously still ready to be amazed. "No one, of course, is championing the idea that there is life on Mars, including me," he said at the time. "But the people who have reacted to the data by saying that there is no life on Mars are doing poor science—they are not looking at the data. And if you look at our data you cannot say there is no life on Mars. In at least two of the biology instruments, the results are positive for biology. Of course, we cannot conclude that we have detected organisms; we are making allowance for the odd chemistry on Mars because the question is so important. However, the data are such that you could drive a biological system through them—I don't mean an elephant but little bitty things."

Horowitz was meanwhile concluding, to his satisfaction, that it would be very hard to drive even little things through his own evidence; he felt he had strong evidence at last that the activity in his instrument was not biological. This argument, like

Levin's, depended on the verification of his previous result. When the data from the subrock experiment, his third cycle at Utopia, came back, the second peak was a count of 7.5, which was negative even if it wasn't as low as the count of 2 he had received on his second cycle at Utopia—the result that had appeared so devastating, though he had been unable to determine whether it was chemical or biological. Now he thought he could determine which it was. Although the amount of water vapor he had used in the earlier experiment was greater than the amount of water under the rock, Oyama's most recent oxygen peak indicated that the amounts of moisture present in the two cases were nonetheless comparable. Indeed, Horowitz's own first peaks, which probably had to do with the amount of water present, were similar for the two experiments.

If all this seemed to prove was that the amount of water in the vapor that had destroyed the reaction in the second cycle was similar to the amount of water under the rock, the conclusion was one Horowitz thought important. Since any life on Mars would require a small amount of water, he argued, it would certainly not be destroyed by amounts of water natural to Mars—and the amount under a rock would have to be considered both small and natural. "So if the reaction is inhibited—or completely stopped—by these minuscule amounts of water, which we know are indigenous to Mars, it's hard to believe it's biological," he said. "It could very well be chemical."

Horowitz had been looking over the results of all the biology experiments and had decided that the chemical argument was not in such bad shape after all. If his two low Utopia readings supported the chemical argument, then his theory—developed after his previous cycle at Utopia—that the water had released oxygen, which either impeded the synthesis of organic compounds in his instrument or destroyed them after they were formed, was very likely correct. Indeed, in all three instruments, when the amount of water was increased (Levin's and Oyama's second injections, his own wet experiments), the activity always dropped. Apparently, the results from the three biology experiments could all be explained by the presence of the same substances—water, oxidants, and oxygen—though these substances behaved differently in each case. For the first time, Horowitz thought the chemical arguments were convincing—though he was not as close as he thought to proving Mars lifeless.

Although Sagan had been remaining relatively quiet (for him), he participated, with some other Viking scientists, in a press conference held in Washington on November 8. Klein, Biemann, and Sagan made remarks to the effect that whereas no one could say conclusively that there was life on Mars, no one could say conclusively that there wasn't. Sagan repeated the points he had made earlier, in private, that the processes the instruments were discovering on Mars at the very least simulated organic processes, and therefore even if the less interesting alternative proved to be correct and the Viking results turned out to be chemical, they would have deep implications for the understanding of the evolution of terrestrial life. Sagan, however, is not one to keep in mind for long the less interesting alternative, attractive as he might make it. "Is it conceivable that *all* the biology experiment results are not biological?" he asked rhetorically, and then answered his own question: "Conceivable, though not likely." A more practiced debater than most of his colleagues, Sagan soon managed to set nearly everyone else's arguments on their heads. For example, with respect to Oyama's experiment, Horowitz had often said it was so wet and terrestrial that, should Oyama detect his Great Event, the burden of proof would be on him to prove that he had not detected a terrestrial stowaway; Sagan, who had been more worried than most about contamination of the instruments by terrestrial microorganisms, now turned the seeming liability of Oyama's instrument into an asset, by saying that his failure to detect life had the advantage of giving the entire biology package a clean bill of health regarding organisms brought from Earth. Moving on to Biemann's negative results, which most people found discouraging, Sagan said that all they really proved was that there was less organic material lying around in the soil on Mars than on Earth. The failure of Oyama's instrument to find signs of life, he felt, was not significant because the instrument was less sensitive than the two that used radioactive labeling. "Therefore," he said, "the negative results for biology are not inconsistent with the positive results."

Indeed, there *had* been positive results, and Sagan was not about to fudge them over, as he had the less favorable ones, with a double negative. "Let me stress that, according to the criteria established before the mission, [Horowitz's and Levin's] experiments have given positive results for microbiology," Sagan said. As the experiments had originally been conceived, affirmative results were presumptive of life—though now the scientists had changed their own criteria for evaluating them. Though Sagan conceded there might have been some reason for the change, he nonetheless managed to leave the impression that scientists who went around changing their criteria after the results came in were laying themselves open to the charge of reinterpreting their data in the light of what they wanted the answer to be.

Having put his audience off balance, Sagan offered to set it straight. "Now I would like to present to you in a very rough manner a couple models for life on Mars which seem to me to be consistent with the data," he said. Before he could do that, though, he had to dispose of the scavenger model Biemann had proposed; Sagan did not seem to appreciate other people's models for Martian organisms. He dismissed the scavenger as that model "in which Martian microbes are extremely efficient at eating the dead bodies of their fellows in an attempt to frustrate Biemann's experiment"—as though, in relation to other proposals for life on Mars, there was something more than usually absurd about the suggestion. "I am not fond of it," he said. "You must imagine [the microorganisms] very efficiently scavenging, so that there is very little organic matter to be found outside of the organisms themselves. This requires a level of effort by the hypothetical Martian organisms that seems to me unlikely to be something that has been described"—that is, whatever the Martian microbes might be, they did not appear to give signs of activity of a level Sagan felt was commensurate with so much effort.

Sagan had begun to think that the most likely model (it was no more than that, he stressed) for life that might exist on Mars and that might create the sorts of results the instruments were getting, was one with a hard shell; he had in mind a kind of iron-rich silicate that would not only be a protection against the ultraviolet light and oxidants, but would also resist high temperatures. While the latter characteristic might not be of much value in the distinctly nontropical climate of Mars, it would be of great importance once the microorganisms got inside the biology package or the GCMS. "Say some of these bugs get shoveled into Biemann's GCMS," Sagan said. "They are heated to five hundred degrees during his pyrolysis. The heat can't break the silicate bonds. Biemann's instrument is full of bugs, and he's not registering any!" In addition, the hard shell—the *right* hard shell—would be a sort of semipermeable membrane that would let moisture in but not out, a useful device in a dry world, for the microorganisms would become traps for whatever moisture was there; however, once they got doused with Oyama's chicken soup, they would absorb the liquid until, unable to eliminate any, they would burst, possibly after a scream for help—Oyama's oxygen peak. The microorganisms would not do much better in Levin's instrument either because that instrument also had too much water or because the microorganisms were not the heterotrophic variety Levin's instrument was designed to detect. Furthermore, the heat-resistant hard shell should surely have enabled the microorganisms to survive Levin's cold, 50° sterilization. All in all, Sagan did not think Levin had detected his microbe.

Sagan's hard-shelled organism, presumably an autotroph, would do much better in Horowitz's dry environment, where it would produce all his active results. (As bugs are heterotrophs, Sagan was using the term *bugs* loosely.) In contrast to Levin's cold sterilization, Horowitz's heat-sterilized control—his second cycle at Chryse, when he had roasted his soil for 3 hours—had produced the officially negative count of 15 that Horowitz himself had admitted might be a sign of some activity. Sagan felt the count was evidence that Horowitz had detected his microorganisms, whose hard shells would have enabled some, at least, to have survived the 160°; indeed, the 15 was about all the activity Sagan expected the microbes would be able to muster after such a sizzling.

At a single stroke, Sagan had turned every single argument upside down, along with the people who had made them: Horowitz, one of Sagan's most severe critics, had provided the best hope for his hard-shelled organisms. Levin, who hoped his results were biological, had not found them. Under the circumstances, therefore, it was not surprising that Levin—who normally might have been expected to look with charity at any model for life on Mars—instinctively tried to squash Sagan's microbe. The hard shell, he said, would confine the organisms inside what amounted to a pressure cooker; while this would not necessarily be a disadvantage in the freezing Martian weather, once they got inside Biemann's, Horowitz's, or his own instrument and were heated, either in sterilization or pyrolysis, any liquid trapped inside them would turn to steam and they would explode, spraying easily detected organics every which way.

"I don't buy it," Levin said.

And Horowitz said of the hard-shelled model: "It's terrible! It's totally uncompelling and ad hoc. It's simply to preserve the notion that life might exist on Mars." On his office door, Horowitz had tacked up a picture of Sagan printed in the same format used for photos from the landers; Sagan, an enigmatic smile on his face, was tinted iron oxide red and stood against a rusty background—evidently, in Horowitz's opinion, as likely a candidate for life on Mars as any other.

Mars was now approaching the point in its orbit where it would disappear behind the Sun. Though Mars would actually be occulted only a few days, communications with the four Viking craft—the two orbiters and the two landers—would fade and then be cut off over a period of about 1 month. Early in November, before communications began to fade, the engineers put the landers into what they called a passive mode. Oyama's instruments in both landers and Levin's instrument at Utopia were left running during the occultation period, for they were in the middle of long incubations; the meteorology instruments in both landers and a seismometer at Utopia were left running, too; the computer was left on and so was the tape recorder that stored scientific and engineering data. Everything else—including the remainder of the biology package, the GCMS, and the XRFS—was shut down. Both landers' sampler arms were retracted into their housings so the scoops would be protected, and the cameras were turned so that the long vertical slits they used in place of lenses were behind protective posts, to guard against windstorms.

Operations closed down largely at J.P.L. too; during the occultation period, most of the scientists and engineers went on vacations—their first, in some cases, in many years. A few of the biologists stuck around, however, for there would be further cycles of experiments after occultation, and new branches had to be grafted onto decision trees. Otherwise, the landers, and most of the scientists, were as dormant as the sporelike microorganisms Sagan and others had once postulated as a likely model for life on Mars. Indeed, after occultation, the Viking program would never return to the frantic pace of the primary mission, the intensive time that had begun with the touchdown of the first lander on July 20; after occultation, when Viking began what was called the *extended mission*, many of the scientists would not return to J.P.L. full time, but rather would go back to their universities or laboratories, receiving their data by mail. The number of flight engineers, too, was cut back; during the extended mission, it would take increasingly longer to send commands up to the spacecraft and longer still to get the data back—3 weeks, in some cases. Nevertheless, in the coming months the question of life on Mars would come much closer to a resolution.

Unlike the dormant Martian spores that theoretically awaited a drop of water to bring them back to life, the landers awaited only a radio signal once Mars was well out from behind the Sun; this came on December 16. Nothing that looked like a sign of life had been recorded in the interval. Because Oyama's and Levin's long incubations would continue for some time, and because Levin was still saving his remaining incubation chamber at Chryse to repeat the cold sterilization, Horowitz was the only biologist in a position to begin a new experiment right after communications were reestablished. Toward the end of December, he started at Chryse, for his fifth cycle there, the sudden-death experiment he had wanted to do earlier. First, he wet his soil sample—one that had been sitting in the biology package's hopper since October—with water vapor, which he was increasingly glad he had sent along, and 4 hours later he heated it to 120° C (well over the boiling point of water on Earth), and then lowered the temperature to 90° for 112 minutes, leaving open a vent to the outside, to drive out the vapor. Then he closed the vent, lowered the temperature to 17°, introduced the radioactive atmosphere, and allowed the incubation to proceed as usual—with the lamp on to encourage activity. Any positive result, he reasoned, would have to be chemical, for organisms don't come back to life after they have been drowned and then dried by extreme heating. A chemical reaction, though, might resume after such exposure to moisture and heat.

Horowitz was not expecting a positive response, for he was still thinking in terms of oxidants, which are destroyed irrevocably by water. The negative response he *was* expecting would not help distinguish between chemistry and biology—at least not directly—but it might make the distinction in an indirect way. He still did not know whether the borderline count of 15 he had received in the heat-sterilized control cycle back at Chryse was in fact a negative or represented a small amount of activity. He had never got an indisputable negative at Chryse, such as he had got at Utopia—and those Utopia results were of no help in evaluating the Chryse instrument. This time, if the sudden death at Chryse produced a low count—a 1 or a 2—he would know that the 15 of the earlier control represented activity, and the activity would be chemical, of course, for that soil had been heat-sterilized to kill any organisms. (Evidently Horowitz did not think it necessary to deal with the possibility of the survival of Sagan's hard-shelled microbes.) Horowitz was obviously proud of what he regarded as an almost foolproof trap: he was in the happy position in which either a positive or a negative result would have to be chemical. Though this sounded like a heads-I-win-tails-you-lose proposition, it in fact was not, for the result might be a count that was somewhere in between. "The only answer that will leave us where we are will be another fifteen," Horowitz said, and for that very reason, a number of his colleagues were betting that was what he would get.

He didn't. But what he did get was almost worse, as far as Horowitz was concerned: a highly active 34—and possibly more, if the instrument's leak was taken into consideration. In any event, it was a clear positive and therefore a strong indication that Horowitz's activity was not biological. More had been killed than organisms, however. Just as the destruction of the activity after his second cycle at Utopia, when he first added vapor, had been consistent with all his chemical theories, the present result, when he added vapor and the activity *wasn't* destroyed, was consistent with none of them—or almost none. The latter result was a reversal of the former. Because he had not only destroyed the oxidants but also got rid of the water (something he had not done before) and had nonetheless received a positive result, neither water nor oxidants could be involved in the active result. That meant that almost none of his original ideas for generating organic compounds under short wavelengths of ultraviolet light (for example, formaldehyde from carbon monoxide plus water) and no variation of them that would work under his filtered light (such as formic acid from carbon monoxide plus hydrogen peroxide) were any longer in the running, because the key ingredients were gone. Even his exchange theory—currently the front-runner because in laboratory tests it had been giving second-peak results in the upper 30s, or higher than any of his other chemical processes—was weakened. Although the exchange got around many of the problems of the other theories by postulating that the organic compounds had been formed *outside* the lander, at least some of those compounds, once they had been scooped into the instrument, should have been destroyed as soon as the vapor released the oxygen. Yet the current result was positive.

The sudden-death experiment had succeeded in knocking

out almost all the chemical arguments. The only one left was the process involving hematites, the stable iron oxides that were the prime candidates for the red coating on the surface of Mars; since hematite cannot be broken down by water, it could survive the wetting to react after the water was gone—when the water would no longer supply a pathway to any oxidants destructive to organic chemicals. The hematite theory was clearly the new front-runner, if only by default. Horowitz still did not know how the hematite—which presumably was a principal cause of Mars's rusty hue—worked. The hematite process, though, posed one difficulty: in repeated laboratory tests (they had been done successfully *with* the filter), it had never given a result higher than the 20s—well below the results he had been getting from the exchange method and a far cry from the 96 he had got during his first cycle at Chryse. He would not believe he had found the right chemical reaction until he succeeded in duplicating that. After the sudden-death experiment, Horowitz felt his results didn't make any sense. He shouldn't have got a positive at all. He felt that all his ideas were about to undergo another major reversal; this was about the fourth time he had had to abandon his favorite theories since Viking 1 landed, and it might not be the last. Perhaps hardest to give up was the elegant simplicity of the idea that the results of his instruments were linked by water and oxidants to the results of the two others.

There was one tenuous way out of the situation: perhaps the water vapor, critical to the sudden death, had never got into the incubation chamber. What made this plausible was that the experiment's first-peak result had been very low, and in both previous experiments in which water had been present—the "wet" experiment and the subrock one, his second and third cycles at Utopia—the first peaks had been very high. The first peaks did not necessarily prove anything; nonetheless, the Chryse instruments had been on Mars for 6 months, and it wouldn't be surprising if there were malfunctions in addition to the leaks he already knew about. On the other hand, since he had heated his chamber to drive out the water vapor before the first-peak measurement was made, the low result was logical whether water had been present originally or not, and there was no way for him to tell which was the case. He decided to use his sixth, and last, incubation at Chryse to see if the instrument's water valve was working properly. However, Horowitz had to wait more than 2 months—until the middle of March—before he could try this: he had to let the other instruments catch up.

Levin, of course, was delighted with the way Horowitz's results further complicated the chemical argument. Serious problems confronted "those proposing the single-cause theory— oxidants—to explain all three instrument results," Levin later implied in an interview in the periodical *Science News*. He seemed to forget that Horowitz's results posed even more serious problems for the biological argument. Levin's own position was not a very comfortable one. His long incubation of the subrock sample at Utopia—to see whether the slow, provocative rise that always followed his second injection would this time take off exponentially—was approaching its ninetieth sol without showing any sign of accelerating in a manner that would confirm the biological indications of the first injection's results. To make matters worse, Oyama—who, with so little going for him on Mars, had been busy in his laboratory at Ames—was about to announce that he had found a chemical explanation for Levin's slow rise. As long as there had been no such explanation, biology could not be eliminated as its cause, even without an exponential acceleration. Oxides and peroxides couldn't account for the slow rise because they would have been used up after Levin's first injection. But now Oyama had discovered in his laboratory that when he poured Levin's nutrient onto maghemite—gamma Fe_2O_3, a magnetic form of hematite—it reacted much more slowly than other oxidants did and at a rate consistent with the release of gas in Levin's long incubations. It was quite likely that maghemite (and possibly also magnetite, another magnetic iron oxide) existed on Mars, for a couple of magnets attached to the scoop appeared to have picked some up. Levin felt that the introduction of maghemite simply added an epicycle to the chemical arguments, further complicating them. Nevertheless, as the lifetime of the instruments drew to a close, and as the time between cycles got longer, the scientists would resort increasingly to their laboratories, and increasingly, their interest would focus on iron oxides.

With the termination of his long incubation of the subrock sample, at the beginning of February, Levin was at last able to start a new experiment at Utopia—his first since occultation. Oyama was still incubating his own subrock sample; Horowitz, however, was ready to start a new experiment himself, so he and Levin did their fourth cycles at Utopia together. They got fresh soil. Horowitz once again tried to duplicate his 96—an attempt he had never made at Utopia—but he couldn't complete the experiment. The valve leak on the way to the detector in that instrument, which had been getting steadily worse, precluded any useful result, and he had to turn his attention to ways of circumventing the leak in the future. Levin was more success-ful; for his fourth cycle, he tried to repeat the cold sterilization which on his second cycle at Utopia had shown that his reaction was greatly suppressed by the relatively low 50° heat—a result presumptive of life. Only the curious ups and downs had prevented acceptance of the data. Now, as winter was approach-ing at Utopia, the instrument's heaters, using the same amount of current as before, could raise the temperature only to 46°; this was enough. As had happened the first time, there was a major reduction in the reaction; the line went only as high as 4,000 before leveling off—a drop of about 70 percent from the normal spectacular curve. What is more, the line held firm, without any ups and downs; it was not far removed from the straight line of Levin's original heat-sterilized control on his second cycle at Chryse. Nobody could question the data this time.

Clearly, Levin—despite the two long, inconclusive incubations and Oyama's maghemite—was very much back in business, and he was delighted. Not only had he got the confirmation, but since the temperature was even lower than before, the results were even more presumptive of life. However, Klein and others once again trotted out hydrogen peroxide—as they had done after the 160° heat-sterilized control, performed in the second cycle at Chryse, and after the 50° cold sterilization—to explain the apparently biological reduction. It appeared to Levin that every time he lowered the temperature of his control, killing the reaction and improving the chances for biology, Klein and the others lowered their estimates of the temperature at which hydrogen peroxide evaporates. The rules, he complained, were apt to change without notice.

Levin found in a textbook that although hydrogen peroxide did indeed evaporate at 50° C, it did so very slowly—at the rate of 1 percent an hour—so that in the 3 hours of the heating the reaction should have dropped by only 3 percent, not by the 70 percent he had just found on Mars. Horowitz thereupon said that the textbook experiment very likely hadn't duplicated the chemistry of that planet; the oxidant on Mars might not be precisely hydrogen peroxide but something slightly different. Failure to duplicate Martian conditions was a common complaint when one scientist invoked a laboratory result unpopular with another. To Levin, this criticism seemed highly "conjectural"—his word of deepest disapproval—because he didn't think Horowitz knew any more about Martian conditions than he did. Other members of the team came to Horowitz's aid; one biologist pointed out that certain iron oxides, such as hematite and magnetite, were known for their ability to destroy hydrogen peroxide, and that this ability could speed up the destruction of hydrogen peroxide on Mars. "I suspect that Levin's experiment on Mars somehow did cause hydrogen peroxide to degrade seventy percent in three hours," this scientist said.

Levin was beginning to feel persecuted by this chemical, which he seriously doubted was present on Mars; he thought he had pretty well eliminated it back in October, at the time of his first cold sterilization. In a final attempt to explode the hydrogen peroxide theory, he irradiated with ultraviolet light a series of possible Martian soils, slightly moistened with water vapor—that being how the oxidants were presumably formed on Mars. Then he put the soil, which was biologically inactive, in a duplicate of his instrument and added his liquid nutrient; when he failed to detect any sign of activity, he took this as evidence that there were no oxidants present either in his laboratory soil or in the soil on Mars that were caused by direct irradiation with ultraviolet light. His critics howled. They said that he had clearly failed to duplicate conditions on Mars and that all he had proved was that he couldn't make oxidants on Earth. Levin, though, was so certain he had eliminated the possibility of ultraviolet-formed oxidants on Mars that he had no intention of experimenting with them further. He said: "We do not believe in peroxides anymore."

Excited as Levin was by the results of his fourth cycle at Utopia, they paled beside the results of his fourth cycle at Chryse. He ran it along with Horowitz's sixth cycle there, which was the control on his sudden-death experiment, to determine whether the water vapor essential to that experiment had been injected properly. Levin did his double-squirt experiment—the one he had been forced to put off until he had been sure of repeating his cold sterilization—in which he injected twice as much nutrient in order to determine whether the critical ingredient was in the soil or in the nutrient. As matters turned out, many things went wrong, and instead of determining the critical ingredient, he wound up accomplishing by accident an even colder sterilization.

Since the soil in the hopper had been sitting there for 140 days—it was now the middle of March, and the last time soil had been collected at Chryse was back in October—both scientists wanted a fresh sample. There was another incentive as well, for the Viking engineers had finally agreed to let the scoop dig a hole a foot deep—something Sagan, among others, had been pushing for. The new sample would be from even farther down than the one taken from beneath the rock at Utopia. It was even more likely that there would be no oxidants and more moisture—perhaps even methanogen-like microbes or, at least, organic compounds. Biemann was not involved this time, for his instrument at Chryse had suffered a short circuit because of cracked insulation and had been shut down lest it damage other equipment. However, he felt satisfied that he had proved conclusively that there were no organic compounds on Mars—not at the two Viking sites, anyway. Horowitz badly wanted the deep-hole soil, but Levin had another idea; he had seen colored patches—greenish, he thought—on some rocks which, in his opinion, could be of biological origin, and he wanted to sample them. He was outvoted by the other members of the team, who failed to see green where Levin did. But then, neither Levin nor Horowitz could accommodate the deep-hole soil, for there was no place to transfer the old soil already in the hopper, and neither of them had any room for surplus. (Once in the hopper, soil either had to stay there or be taken into the instruments—there was no other way out.) Accordingly, both scientists cast longing looks at Oyama's instrument: it would make a fine dump. Oyama's instrument, as it happened, had no more room in its single large dump cell either; if it were to accommodate the overflow, it would have to take dumped soil directly into its incubation chamber, thereby ending the instrument's life as an experiment. The situation was reminiscent of the time Horowitz had wanted to go ahead with his cold incubation at the others' expense—only this time he had Levin on his side; Levin didn't believe in Oyama's oxidants any more than Horowitz believed in a Great Event. Oyama couldn't resist the two of them; he gave in, stipulating only that he be allowed to finish his 200-day incubation, which still had a couple of weeks to go.

As it happened, Oyama did better out of all this than either of the scientists who were willing to throw him overboard—and especially Horowitz, whose plans had a way of backfiring. The digging was done in a barren-looking spot that the geologists, with possible ironic intent, had called Atlantic City, but which everyone now hoped would be crawling with as much life as its terrestrial namesake. The ground at Chryse was virtually unaltered since the last trench had been dug in October, though there was one sign of activity—a slight slumping of soil in a dune. (It caused a flurry of excitement among the geologists. "It may not sound like much, but it shows the scene is dynamic, the landscape is changing," one member of the imaging team said. After the earlier hopes for activity on Mars—erupting volcanoes, giant Marsquakes, rushing rivers, swirling dust storms, perhaps even the ground trembling under the feet of giant macrobes—it was a distinct lowering of expectations.) Digging the trench took about a month, and though everything was quiet, Levin and Horowitz, already impatient, became alarmed when, before the sample was taken, there was a problem with the radio receiver at Chryse which threatened the engineers' control over the Viking 1 lander. Horowitz and Levin panicked; rather than risk waiting any longer, they elected to go ahead with the next cycle, using the soil that had been in the hopper since October. Oyama was reprieved: he would not have to sacrifice his instrument after all. The two others were naturally disappointed—the more so when, after they started their incubations, the radio problem cleared up, for this meant that they could have safely waited for the sample from the deep hole. As it turned out, though, Levin was lucky to be using the old soil.

Levin and Straat injected the two squirts separately, with a 3-hour interval between injections. "We needed a control over whether the hopper had done something to the soil in that time or not," Straat explained. "We knew what the reaction to a single shot should be—the usual spectacular count of twelve thousand to fifteen thousand per minute. But if the shots were simultaneous, how would we know that the result was not due to the long time the soil was in the hopper? With two variables, we couldn't separate them—unless we did two injections." For whatever reason, something different happened this time. There was no spectacular rise after the first injection, as there should have been, nor was there the usual drop after the second; rather, after the first injection the count rose slowly to about 400 a minute. After the second injection, it leveled off for a while and then started to climb again, reaching about 2,000 by the end of the cycle—after 7 sols. This was very little. Clearly, the active, or limiting, factor in this cycle was not a shortage of nutrient; more important, though, something had caused the active agent to degrade, and Levin and Straat didn't have to look far for an explanation that satisfied them, at least. The temperature in the hopper had all along been averaging 15° C—somewhat higher than the highest Martian summer temperature at that site.

"The fact that the soil spent one hundred and forty days in the hopper at fifteen degrees is the only possible explanation for cutting down the activity," Straat, who thought the result favored biology, said afterward.

"Our group's case for biology has consistently strengthened

since the first day," Levin said at this time. "Every result we have from our instrument is internally consistent and consistent with life." Consistency, he felt, was what the chemical argument lacked.

The rest of the biology team was not impressed by Levin's interpretation. Klein and others suggested that peroxides, too, might degrade at 15°—especially over a 4-month period.

In science, the struggle for survival of ideas or theories is at least as persistent as the struggle for a livelihood is among microorganisms. Quite clearly, as men or their machines go off into space, one hallmark of our evolution they will take with them is a continual combativeness and argumentiveness. Very likely any forms of life Earthlings encounter on other planets will not find this characteristic so strange, for they too will be products of evolution, and therefore they will almost certainly have undergone similar struggles for survival themselves; a tendency for dispute may well prove to be a universal characteristic of life. Klein, Horowitz, Oyama, Levin, Sagan—and the rest of humanity—would probably feel at home, in this one particular, in whatever odd corner of the cosmos they or we could imaginably turn up.

With his newly reprieved instrument at Chryse, Oyama wanted to try something different. Because his 200-day incubation had ended without a trace of a Great Event, he had now given up the idea of life, and he had at last decided to use his instrument at Chryse as a chemical laboratory—a decision Horowitz, who had urged it many months before, said was better late than never. Horowitz was even glad now that Oyama's instrument hadn't been sacrificed. Completely ignoring Levin's protests that there were no oxidants on Mars, Oyama, whose belief in their presence ran deep, wanted to get at the question of whether there might be *more* than one.

Accordingly, at the same time Levin was doing his double squirt, Oyama heated his soil to 145° C; most superoxides are stable at temperatures up to 200°, while most peroxides break down at temperatures of 100° or less, and he had picked a temperature right in between. Then he humidified the soil and analyzed the atmosphere again; he got an oxygen peak, presumably caused by the destruction of superoxides that hadn't been destroyed by the heat. Thus, Klein may have been right in something he had said at the time of Levin's heat-sterilized control, his second cycle at Chryse, when under circumstances remarkably similar to Oyama's present experiment, Levin had heated his soil to 160° and then added liquid nutrient; when he had got no reaction, Klein had suggested the superoxides were present but for some reason were unable to oxidize Levin's nutrient. Klein had been quite certain, though, that the degradation of activity in Levin's control could be explained by peroxides. Indeed, Oyama now felt that the result of Levin's experiment, taken together with the result of his present one, clinched the matter of the presence on Mars of both peroxides and superoxides, but Klein, though he felt the evidence was persuasive, cautioned that it was still circumstantial.

Most people on the biology team saw no reason to stop at just a couple of oxidants in a highly oxidized environment like Mars. They thought that there could be several varieties of peroxides and several varieties of superoxides, each with different characteristics, and that modeling the correct mixture to account exactly for all Levin's and Oyama's results could take years of laboratory work. Just the rough model they had now was complicated enough: it seemed that hydrogen peroxide accounted for all Levin's results—the activity at normal temperatures and the cessation of activity at higher ones; that a combination of peroxides and especially superoxides accounted for Oyama's results at normal temperatures; and that superoxides alone accounted for his activity at high temperatures, such as the oxygen peak he had just detected at Chryse. And there was also the maghemite gamma Fe_2O_3, which Oyama had found might account for the long, slow rise in Levin's count. Yet despite the complexity of the soil—it seemed almost alive itself—none of these compounds were able to account for Horowitz's results. Oyama, having worked so hard to try to explain what was going on in his own and Levin's instruments, wouldn't rest until he had done the same for Horowitz's.

Horowitz's sixth and last cycle at Chryse, which he did along with Oyama's chemistry experiment and Levin's double squirt, was his long-postponed control on his fifth cycle there—the sudden death, whose high second-peak result had appeared to rule out almost everything. But had the water, which should have killed the reaction, actually got into the chamber? Because the first peak had been unusually low for an experiment in which water was used, the possibility existed that the water valve had stuck, and there had been no water. To find out, Horowitz now wet the soil with a double squirt of vapor, and this time did not heat the soil afterward, so the water would not be driven off. This was almost a duplicate of his second cycle at Utopia, in which he had learned that water killed his activity, and if everything was working properly he should get the same results now. (Because of the leak, he had to put in many times the usual amount of radioactive gases to make sure the right quantity got in. "You see how close to the ragged edge we were with the instruments," Klein said one day in his office at Ames.) The results, however, were contradictory. The first-peak result was very high, indicating that the vapor had in fact entered the chamber, but the second-peak count, instead of being near zero, was 33—well within the active range.

Klein called the 33 a shocker. He had been keeping track of all Horowitz's second-peak results, and he and others felt that at last, taken together, those results were beginning to fall into a pattern:

Chryse 1—light, dry, active
(the original one)   96
Chryse 2—light, dry, sterilized
(the heated control)   15
Chryse 3—light, dry, active
(attempt to duplicate original)   27
Chryse 4—light, dry, active
(attempt to duplicate original)   35
Chryse 5—light, wet, sterilized
(sudden death)   34
Chryse 6—light, wet, active
(control for water)   33

The last four results, Klein and others noticed, were close, ranging from 27 to 35. The first two results—the original 96 and the 15 of the heat-sterilized control—were special cases, they now felt, which could be ignored. If the last four results were the only important ones, it didn't seem to matter much what was done to the soil: whether it was wet or dry, hot or cold, the results were similar. And the same could be said of the results at Utopia, where, since there had been no heat-sterilized control, the second experiment did not have to be eliminated and so could be counted:

Utopia 1—dark, dry, active   23.0
Utopia 2—light, wet, active
(first wet cycle)   2.8
Utopia 3—dark, dry, active
(subrock, wet)   7.5

Here, if the result of the first cycle was removed, the

remaining experiments also gave results that were close together. Two experiments, of course, were too few to indicate trends, but Klein and some of his colleagues pointed out some nonetheless; they had little choice. From the Utopia data, they were tempted to think that whether the experiments were run in the dark or in the light made no difference: though the first experiment at Utopia had been run in the dark (this had been done out of necessity, because of high temperatures after landing), it had given an active result; the third one, also run in the dark, had given a negative result. The Utopia second peaks, of course, were much lower than the ones from Chryse. "We had convinced ourselves that this was because there was more water at Utopia—either because we ran the experiments wet or because the soil had more water in it," Klein said, sitting at his desk. "However, at the time the wet experiments were being done at Utopia, there had never been a wet experiment at Chryse." Later, when the two wet experiments were done at Chryse, the results were quite high. Klein and others were now tempted to ascribe the higher results at Chryse and the lower results at Utopia to discrepancies between the two instruments. "It now looks as though the water, the light, the heat, and everything else we did, regardless of which lander we used, made no difference," Klein said.

Before this theory of a uniformity in Horowitz's results—which was still only a theory—could be accepted, the results of his first cycle at each site, which were by far the highest for each instrument, and were therefore out of step with all the others, had to be eliminated. One thing that had bothered almost everyone about Horowitz's instrument even long before the mission started was the possibility that the affinity of the organic vapor trap—the golden U-shaped pipe that the gases passed through on their way to the radioactivity detector—for carbon dioxide was greatly enhanced by long periods in vacuum. During the 11 months it took the two Viking landers to reach

Mars, the valves in the biology instruments were open to space, so before the launch the organic vapor traps of both instruments had been saturated with ordinary, nonradioactive carbon dioxide (after the landings the traps were again saturated with the Martian atmosphere, which is 95 percent composed of that gas) to prevent the traps from retaining excessive amounts of radioactive carbon dioxide later. "But suppose the eleven-month exposure to vacuum enhanced the trap's ability to pick up carbon dioxide far beyond our estimates?" Klein asked at Ames. "And suppose we didn't saturate the traps sufficiently? And suppose each trap then picked up some radioactive carbon dioxide from the chamber's atmosphere during the first peak of the first cycle, which was later baked out by the pyrolysis for the second peak? That would say that the first time you do the experiment, you will get an immensely high second peak. And that's the way the data on both instruments look." If this was so, then it was little wonder that Horowitz had never been able to duplicate his 96 on Mars.

Horowitz did not like any of this—despite the fact that, at the very beginning, he himself would have welcomed a way of eliminating the 96. Now he seemed almost defensively proud of it—the way Oyama was of his oxygen peak and Levin of his curve. He thought it his most significant result, but later he came to think of it as an anomaly (but not an instrument anomaly) whose duplication in the laboratory would be an important step in proving the chemical theory. This was Klein's and Horowitz's only serious disagreement. Horowitz pointed out that he himself had been the one to discover the effect of vacuum on his organic vapor traps and said that, although he had never tested a trap in such a strong vacuum for as long as 11 months, he did not think even that length of time in space would account for his 96. As for discrepancies between the Chryse and the Utopia instruments, although Horowitz had never been able to test in the laboratory the instruments now on

Mars, he had run many tests with an identical instrument, and it had given more or less the same results from the same soils every time, and this suggested that the instruments on Mars would, too. Now, with his theories once again in disarray, Horowitz clung rigorously to the integrity of his instruments and of his results—including the two initial ones. He recognized that it was difficult to make much sense out of his results when they were taken together, but he preferred another explanation to Klein's. Perhaps none of the changes he had introduced *had* affected things much (itself an interesting piece of information), but he himself was inclined to believe that the differences between the Chryse and the Utopia results were caused not by discrepancies between the instruments but by differences between the soils at the two sites—particularly with respect to the effect of water vapor upon them. Perhaps at the time of the first cycle at Utopia, when he had been unable to turn his lamp on, he had been too quick to accept Levin's and Oyama's evidence that the soils from the two sites were similar. "There could be a heterogeneity of the soils such that at Utopia water kills my reaction and at Chryse it doesn't. That's the trouble when you can't see what you're doing; and that's about all the data we have," Horowitz said in his office at the California Institute of Technology, echoing an earlier complaint of Biemann's. The game of blindman's buff with instruments was almost over, perhaps with no winners. Klein admitted that Horowitz's explanation could be right. "Perhaps the soils at the two sites *are* different," he said. "Or perhaps the instruments are different. At this point, I don't see how we can unravel that one."

Oyama, after his success with the oxidants, had been spending 15, and even 20, hours a day in his laboratory at Ames trying to come up with an explanation for Horowitz's results. Now he proposed that if the superoxides were doing the dirty work in his own instrument and the peroxides were doing it in

Levin's, possibly Horowitz's results could be explained by another type of chemical—the carbon suboxides. This notion had a certain elegance: suboxides, being compounds that contain less oxygen than the oxides, were off the bottom of the ozonide–superoxide–peroxide–oxide ladder. (As some of them are reddish yellow and turn green when exposed to moisture, they had long ago been proposed as an explanation for the seasonal changes on Mars.) But Horowitz didn't like this idea. He didn't believe there was any carbon suboxide on Mars. In this, Horowitz was supported by Biemann, who was certain that if any was present his GCMS would have found organic compounds derived from it—and it hadn't.

When in his next cycle at Utopia—another chemical experiment—Oyama claimed that he had detected evidence of a carbon suboxide on Mars, and further, that it might be ancestral to organic chemicals, Horowitz blew up. "There's no reason to give any more weight to Oyama's hypothesis than to the ones we're studying—perhaps less," he said. "I have the feeling Oyama is sort of pulling something out of the air; it's terribly ad hoc. Mariner 9 had an instrument to look specifically for carbon suboxide and found none. None has been identified by Viking, either."

Suboxides or no suboxides, Horowitz appeared to be in trouble. After the doubts cast on all his theories by the sudden-death experiment, the further possibility that none of the changes he introduced had made any difference appeared to leave Horowitz—who had initially been the biologist surest of his experiment—without a leg to stand on. This was not the case, however. Despite all the reversals—in fact, because of them—Horowitz had come out on top with respect to the only question his instrument or either of the other instruments had been designed to answer: whether or not there was life on Mars. Whatever the explanation for the activity he had been measuring, the evidence was now overwhelming that—in his instrument, at least—it had to be chemical activity of some sort. If nothing—temperature, light, water—affected his activity, it had to be insensitive, and only chemistry filled the bill. (This would be true if he never managed to duplicate his 96 chemically; if he did, that, of course, would be further proof.) To expect any of the instruments to identify a particular chemical reaction—the task Levin now seemed to be requiring of them—might be asking too much of experiments designed to answer simply yes or no to the question of biology. At the very least, Horowitz's failure to pinpoint the reaction could hardly be taken as a reason not to rule out life as a cause of his results.

Previously, Klein and others had felt that Horowitz's activity was the most difficult to explain chemically, Oyama's the easiest, and Levin's somewhere in between. Now Klein was prepared to put Horowitz's results firmly in what he called "the nonbiological column." Doing so gave Klein himself a powerful push away from the overall possibility of life on Mars, for in his view the only instrument that left this possibility open at all was Levin's, and Klein was skeptical about that. Though Levin's arguments were plausible when his experiment was regarded in isolation, they lost much of their force when they were regarded not only in relation to the results of the other biology instruments and the GCMS but also in the context of the Martian environment as a whole—its low temperature and pressure, its aridity, and its very likely toxic soil.

Despite the conflicts of the last 8 months, Klein's opinion still acted as a barometer of the biologists' thinking. Horowitz, however, was not yet ready to go quite as far as Klein about his own results. "My feeling is that my instrument has not found a biological response, but I still don't have a chemical proof, and as long as I don't have one there will be a residue of doubt," he said. To the end, Horowitz was determined not to yield to his preconceived ideas; his commitment to the scientific method put him in the curious position of leaning over backward so far that—along with Levin—he was among the last to keep the door open to the idea of life on Mars.

In the middle of March, the biology instruments on Mars were nearing the end of their lifetimes—particularly the instruments at Chryse, which had been used more. Horowitz's instrument was through; after six successful experiments, it had all the soil it could take, and as a result of its leaky tank, it was virtually out of radioactive gas. Levin's Chryse instrument, after four successful cycles, had a surfeit of soil too; furthermore, it was virtually out of nutrient because of the double squirt. The only instrument at Chryse that was still working was Oyama's, and for a final experiment he received fresh soil from the bottom of the deep hole. If he had a sense that justice was being done as he was the only one to get the soil for which the two others would gladly have sacrificed his instrument, he didn't say so. Anyway, his own instrument either developed a leak or received too much soil to permit the incubation chamber to be sealed properly, and Oyama received no further results from Chryse either.

The instruments at Utopia were still working, even though the temperature there had been dropping rapidly with the approach of winter in the northern hemisphere. In the south, where it was summer now, the carbon dioxide ice cap had long since melted, leaving only a small residual cap that was probably frozen water; but in the north, carbon dioxide (some of it doubtless from the south pole) had for some time been precipitating back onto the cap. As far back as January, carbon dioxide clouds had begun swirling over the northern cap; even at Utopia, near the fiftieth parallel, the light dimmed, shadows disappeared, and the temperature dropped. Then early in February, far down in the southern hemisphere, huge clouds of dust sprang up—dust devils, presumably, stirred up by winds rushing down the slopes of some of the big basins near that pole. The dust was caught up by global winds high in the atmosphere, and by the end of February most of the planet was shrouded for several days, as it had been in 1971 at the approach of Mariner 9. Down on the surface, the storm was not as severe as it looked from above. (In dust storms, particles fill the atmosphere, making it opaque when seen from space, but they do not limit visibility much on the ground. Also contrary to expectations, the landers never recorded wind velocities during storms much in excess of 60 miles per hour, though in the area where the dust was raised near the south pole, the winds might have reached a speed somewhat in excess of 100 miles per hour—far short of certain pre-Viking estimates.) At Chryse, the landscape darkened a little, some dust landed on the camera, and the temperature plunged 20° in a single day. By early April, Utopia was so cold that the GCMS on the Viking 2 lander had to be turned off permanently for fear the cold would crack its electrical insulation. Both of Biemann's instruments had now done their work. (If Biemann, the first to lose both his instruments, felt victimized by circumstance once again, he didn't say so.) At night, the temperature at Utopia reached the precipitation point of carbon dioxide: $-123°$ C. In photographs taken in the early morning, before the Sun had a chance to melt them, patches of white—ice, presumably—dotted the landscape.

Clearly, conditions were right for Horowitz's cold incubation—the one he had tried to do in November, even though, because of their liquid nutrients, it would have then meant risking both Levin's and Oyama's instruments. This time, Levin and Oyama decided to join Horowitz in the cold incubation, not so much out of amiability as out of inevitability. They all thought this was the end. It wasn't quite, however. While the sampler arm was delivering fresh soil, it stuck; the cold, apparently, was too much for it. As it had before, the little robot on Mars just kept chugging on, so there was no way to keep the heaters inside the biology package from turning off. The temperatures inside plunged. Levin and Oyama wrote off their instruments, and as matters turned out, Horowitz's instrument, with its long-term leak, was now pretty useless. It would have returned no data. Yet Levin's and Oyama's instruments, against everyone's expectations, survived the deep cold. Each of the two biologists went on to do one more active cycle, though they had to use soil from the long-held batch and put it on top of used soil in their chambers, and this meant that the results were open to question. Still, Oyama claimed to have found further evidence for his carbon suboxide, and Levin, who detected virtually no reaction, claimed that the lack favored biology, for anything alive would surely be dead after so much time inside Viking. Horowitz, who had already retired to his laboratory, was able to shrug off this final series of outrages.

The biology instruments at Utopia were shut down for good on May 28; the ones at Chryse were shut down 2 days later. "If life on Mars had been at all easy to find, we'd have found it," one biologist said by way of obituary. Certainly the instruments had worked better than anyone had dared to predict. The NASA press release announcing the shutdown stated: "Biologists have not reached any final conclusions about the presence or absence of life on Mars." In the following months, though, the residue of doubt dwindled further. Since the hematite theory had been about the only survivor of the sudden-death experiment, Horowitz went on to experiment with other iron oxides that might exist on Mars; he found that by using hematite, maghemite, and magnetite, he was able to get results similar to his Martian ones. Though the chemistry of the reactions was not clear, the hematite had apparently needed the filtered light to synthesize organics, and the two others had not—another mark in their favor, considering that in his first cycle at Utopia Horowitz had got a positive response with the lamp off. He thought it likely, though, that all his Viking results could be explained by some kind of iron chemistry. Other chemical theories were also advanced, including one that attempted to explain the partial duplication in laboratories of the Martian pyrolytic-release and label-release data, using certain iron-containing clays, among them a form of montmorillonite enriched in iron—a theory not inconsistent with Horowitz's new line of thought. By September, however, most scientists felt that they were pushing their theories beyond what the Viking data could support—something they had begun to feel when Oyama suggested the suboxides, if not before—and that the remaining doubts might not be resolved without another landing on the planet.

At around this time, a number of scientists, including Horowitz, began speculating about what effect a barren Mars would have on their view of the chemical origins of life and, in particular, on the existence of life elsewhere in the universe. "Getting life started on a planet may be more difficult than we thought before the landings," said one Viking scientist, Tobias Owen, of the State University of New York at Stonybrook, who had been studying the atmosphere of Mars in relation to life. "A sort of pseudo religion has grown up around the concept that every time you get carbon and hydrogen on a planet you get life. However, there may be a good many more factors than we know even now, and it's asking a lot that you find them very often. Mars may be a cautionary tale."

Indeed, a small number of scientists—astronomers not connected with Viking, one at the University of Colorado, another at the University of California at Los Angeles—were beginning to raise the unsettling, not to say unpopular, notion that the number of factors might be so great that life on Earth might, after all, be unique. Horowitz was becoming surer and surer that one of the factors required for life was probably an ocean. "If Mars is barren, it will force us to take another look at the Earth," he said. "It will make our ocean seem more important. On a planet like Mars, you might have all the chemicals you need for life, but they won't meet. Our ocean is a nursery; it provides the solvent—water—in which chemicals diffuse and collide with each other. It's a nursery in that it has a constant temperature over a long period; it provides continuity. Nothing on Earth is as old as the ocean—not the mountains, not even the continents. We have the only ocean in the whole solar system. If, in the end, it turns out that Mars is lifeless, the idea of the ocean as the nursery of life will take on a new importance. Planets without oceans become less likely habitats for life. But in other solar systems there must be hundreds of thousands of other planets with oceans."

Whatever the case, it is quite likely that if ever again mankind receives what might be interpreted as a sign of extraterrestrial life—data from a future Mars lander that could rove the planet, perhaps, or even a radio signal that might be from another intelligent civilization—matters could well follow the same course as they did with Viking: hopeful excitement, doubt, arguments, and confusion, all saying as much about terrestrial as extraterrestrial life and all played against a background of reams and reams of seemingly conflicting computerized data.

Epilogue

Over the next few years, laboratory experiments made a chemical explanation for the results from the instruments on Mars look more and more likely. In March 1979, Lederberg, at The Rockefeller University, confirmed that he now definitely associated himself with the generally chemical views expressed by Klein. And Horowitz, using iron-rich clays, has finally come so close to duplicating his 96, even perhaps surpassing it, that he now says he is ready to exclude altogether the idea of life currently existing on Mars. The clays have gotten good results in the other instruments, too, though to date they have not managed to duplicate the sensitivity to temperatures found in Levin's results. The oxidants are looking more and more likely as the explanation for Oyama's and Levin's results. Levin, despite his vows that he would never experiment with them, nevertheless has done so, for he announced in January 1979 that he has come close to duplicating his cold-sterilization results using hydrogen peroxide and maghemite, getting a

241

negative—as Oyama had said he would. Later, however, he partially retracted; among his second thoughts was that the amount of hydrogen peroxide he had used was about 1,000 times more than could possibly exist on Mars—though, of course, Levin's estimate of what that amount might be is vastly lower than his colleagues'.

Despite everything, there are still scientists who continue to keep open the possibility of life on Mars. "Even if we duplicate all the results in the laboratory," one member of the biology team said, upon hearing that Horowitz felt he was close to reproducing his 96, "that doesn't prove that that's what's going on on Mars." Sagan, never one to appreciate being on the receiving end of a "cautionary tale," as Tobias Owen had called the failure to find life on Mars, said in the spring of 1977, and confirmed in the spring of 1978: "I haven't seen anything that makes me go negative on the idea of microbes on Mars; on the contrary, I'd have to say I stand about where I did before." In the spring of 1979, he repeated that sentiment, adding that of course he had seen nothing that had convinced him that there *was* life on Mars, either, and making a final plea in behalf of the need in science to keep more than one possibility in mind. In 1977, he received a small grant from NASA to examine thousands of Viking orbiter photographs of Mars to look for traces of advanced civilizations, ancient or modern, if only to rule out these eventualities; NASA would look pretty stupid, he argued, if after all the searching for microbes, it missed something like a Great Wall of China on Mars. Although he has set some students to work poring over the photographs, no Great Wall, Inca road system, or Roman viaduct has turned up—something that has surprised no one, including, of course, Sagan. The search is not yet complete.

Occasionally, there is some ancillary development to bolster the hopes for life on Mars, such as the likelihood, announced at a symposium of Viking scientists and others interested in Mars, held at J.P.L. in January 1979, that the pressure on the planet's surface early in its history may well have been as much as 800 millibars, or 80 percent the present pressure at sea level on Earth, enough not only to allow entire *oceans* of water, but also possibly to retain substantial warmth from the Sun. Further, Viking orbiter photographs of the north polar cap in summer, which have not yet been completely processed, make it look as though there are indeed deposits of ice many *kilometers* thick, if not in the permanent, summertime cap itself, then beneath it, interspersed with layers of dust—the circumpolar laminations. In other words, there most likely is a considerable reservoir of atmospheric gases, including water, frozen out at the poles, as Sagan once argued. (Some of this water, Sagan argues more recently, might from time to time be liquid.) This large reservoir does not revive the theory of relatively recent periods of warmth, high pressure, and rainfall that Sagan once argued for; the latest analysis of the Viking orbiter photographs has tended to push the tributary channels—those that have arisen only as the result of the drainage of a broad area after rainfalls—even further back in time, into the first billion years of Martian history, when the atmosphere was presumed to have been thicker anyway.

There was a development on Earth not long ago that caused a flicker of hope among the diehards. In February 1978, it was announced that in the Antarctic valleys—those most Martian of terrestrial habitats—organisms known as endoliths had been found *inside* certain translucent, porous rocks, forming a dark greenish layer a few millimeters deep where the warmth of the Sun is retained; they are blue-green algae, indistinguishable from any other, that have adapted to this particular microenvironment. (The presence of endoliths has no bearing on the old argument between Horowitz and Vishniac about whether organisms can exist in the dry Antarctic valley environment; the endoliths have found their own indoor greenhouses to escape from it.) The possibility had been raised, however, that if life got a start in the earliest days on Mars (blue-green algae are found in terrestrial fossils over 3.5 billion years old), perhaps as conditions worsened some forms of life found a refuge in Martian rocks, of which there are certainly a great many. In the opinion of Horowitz and other biologists, this notion, like so many others, founders on the dryness of Mars, even as compared with an Antarctic valley. The outstanding thing about life on Earth, of course, is the amount of life that exists *outside* rocks and that doubtless provides a continuing vigorous evolutionary mainstem from which that particular tendril, the endolith, was able to push into its bizarre niche.

Late in 1978, Levin published in a British biological journal the news that further analysis of the photographs from Chryse had shown that what he had at first thought were green patches on the rocks were not only green, but they *moved*. "A combination of wind movement of dust and dirt dropped by sampler arm operations could have produced the slight changes in patterns and position," he wrote, covering himself before he started. "However, the observed patches, patterns, and changes could also be attributable to biological activity." The matter is extremely technical optically, as a few pixils one way or another can make a great difference in shades of color—as happened originally, when scientists thought the Martian sky was blue instead of red. However, Levin's Viking colleagues—including technical experts on the lander imaging team—see no reason to think the patches in question are anything other than reddish orange, like the rest of the dust. In addition to dust being blown about by the wind, the movement of the patches could be accounted for by the different Sun angles in the photographs Levin considered; there are many other examples of this sort of "movement" among the pictures. Possibly, the shifting green patches that Lowell and others a century ago saw as covering much of the planet have dwindled to a few evanescent, enticing spots of doubtful hue.

In contrast to a few years ago, the interest among scientists today in the proposition that there might be life on Mars is virtually nil, and within the scientific community this is usually the way ideas die. NASA has so far failed to fund any follow-up mission to Mars, despite the new prospect of cracking open rocks to find microorganisms. The big fluctuations Horowitz described as putting the chances for life on Mars up or down by an order of magnitude have now quivered to a virtual standstill, resulting in a straight line at the bottom, or inactive side, of the graph—like the data from a heat-sterilized control. In evolutionary terms, the idea that life currently exists on Mars has quietly gone the way of the Great Auk. If there is no life on the one planet in the solar system which, next to Earth, had long been considered the most likely to foster it, then the search will have to shift where neither we nor our machines can easily follow. And even out there, beyond the solar system, the waves are dying down, too. In the autumn of 1979, at a convention of exobiologists held in College Park, Maryland, and entitled "Where Are They? A Symposium on the Implications of the Failure to Observe Extraterrestrials," most of the speakers were discouraging about life, intelligent or otherwise, elsewhere in the galaxy. Many of them invoked the Viking results as influencing their opinion. (Sagan and his Cornell colleague Frank Drake did not attend.) Whatever the case, Lander 1 is still sending back photographs from Mars, and it has been programmed to do so into the 1990s. If in the next decade a squamous purple ovoid or anything else should wander slowly enough across the blasted, desiccated Marsscape, we will know.

Index